# CHASING THE SIXPENCE: THE LIVES OF BRADFORD MILL FOLK

Rachel Bellerby

Fort Publishing Ltd

First published in 2005 by Fort Publishing Ltd, Old Belmont
House, 12 Robsland Avenue, Ayr, KA7 2RW

Cover illustration by Andy Bridge

Graphic design by Mark Blackadder

Photographs courtesy of Bradford Industrial Museum
unless otherwise stated

Typeset by Senga Fairgrieve

Printed by Bell and Bain Ltd, Glasgow

ISBN: 0-9547431-8-0

# the author

Rachel Bellerby is a freelance journalist, specialising in the history of West Yorkshire. She was born in Bradford and is proud to say that at least five generations of her family have been born in the city. She studied medieval history at University of Wales, Lampeter and has written over eighty articles for various UK publications, including *BBC History Magazine, Ancestors, Your Family Tree* and *Practical Family History.* Rachel lives near Ilkley with her husband and their young daughter.

*To my parents and grandparents,*
*who taught me to love Bradford.*

# contents

# preface

The words Bradford and textiles go together like warp and weft. Although many of Bradford's textile mills have now closed forever, interest in the lives of those who worked there has never been greater. My fascination with textile mills began when I was told the old Bradford tale that it is possible to drive a horse and carriage around the top of Lister's chimney. I must have been about seven years' old and can remember clearly gazing up at the mill and thinking it must be the most beautiful place in the world.

I have since been told that the tale cannot be true because the mouth of the chimney is far too narrow. Nevertheless, for me, the magic remains. Like many children raised in the 1970s I was brought up on 'mill tales', and loved to hear about my grandparents' lives in a Bradford that was a very different place to the city we see today.

When I placed a letter in Bradford's best-selling newspaper, the *Telegraph & Argus*, asking former mill workers to get in touch, I was hoping for a good number of replies from my fellow Bradfordians. But nothing could have prepared me for the warm-hearted response. Dozens of people got in touch and I spent many enjoyable hours chatting about a way of life that I found I knew very little about. Any romantic or rose-tinted notions I might have had about mill life were soon dispelled and replaced with the much more interesting reality.

The book's title, *Chasing the Sixpence*, comes from an old saying: it refers to those workers who moved from one mill to another looking for better pay. The ease with which people could leave a mill and move to another the same day – even after being sacked – is alien to me. But it was part of the way of life in the mills when times were good. Workers were in such demand that they could more or less pick their place of work.

The chapters that follow cover all aspects of mill life: from living around the mill, days out and holidays, to the working conditions and way of life in a textile mill. Of course, there are tales of hardship and even despair, but all those interviewed had one thing in common; a sense of humour and a cheerful outlook, qualities that I feel many Bradford people share.

There are many excellent books about the history of the mills and the production processes. This book doesn't attempt to replicate them. There is information about the textile trade here, but I hope that I have given just enough information to complement the mill workers' stories.

A book like this could not have been written without the involvement and goodwill of those who were interviewed. These were the people who lived through a period that is a crucial part of Bradford's history. I am extremely grateful to everyone who took the time to answer my questions and sometimes dig into the deepest recesses of their minds about events that took place many years ago.

And now all these stories have been written down for everyone to discover. They tell of a way of life that has gone from Bradford forever, of friendships that lasted a lifetime and of years of hard work and good fun. Mill work was a specialised and sometimes unpredictable trade and there are many fascinating tales that I have been privileged to share. Now I hope you will enjoy those tales too.

My biggest debt is to all who so generously shared their time and memories. I will never forget the enjoyable conversations I had with Bradford's former mill workers, who are some of the friendliest people I have ever met.

Thank you also to Eugene Nicholson, David Briggs and to Derek Hudson and all at *Writers' News* for their interest and support. The final thanks go to my parents and husband who have endured months of obsessive 'mill talk'. I hope they will not be too fed up of mill life to read this book!

## The contributors:

Saiyid Abid, Kim Andrews, Veronica Bennett, Ernie Blake, Leslie Boocock, David Briggs, Doreen Brook, Molly Carter, Pauline Chapman, Mabel Clements, Kath Clough, Roy Conway, Doreen Craven, Diana Dean, Julie Dillon, Douglas Downey, Bill Edmondson, Brian Fraser, Harry Fuller, Elizabeth Graham, Celia Greenwood, Frederick Hanney, Colin Hardy, Nasreen Hasaan, Evelyn Hindle, Muriel Hodgson, Joan Holmes, Emily Hoyle, Ronald Hoyle, Violet Hurd, Christopher Ingleby, Amel Khan, Marya Krol, Norman Lister, Fahim Massood, Jean Mortimer, Kenneth Nelson, Hannah Newton, Gladys O'Melia, John Pashley, Evelyn Pearson, Maria Pick, Andrea Ratcliffe, Dominica Richmond, Edna Richmond, John Richmond, Diana Roberts, Dee Rogers, Howard Rudge, Ernest Ryan, Kathleen Shuttleworth, Mr Shuttleworth, Hilary Simpson, Derek Spencer, Jean Squires, Edward Stanners, Miss Watts, Khalida Wazir, Winnie Wilcox, Kathleen Wright, Marlene Young.

Rachel Bellerby, Bradford, August 2005

# Introduction

When the first steam engine began operating in a Bradford mill in 1798, the town already had a textile industry stretching back some six hundred years. For centuries, the woollen and worsted industries in the West Riding area had operated through the labour of home-workers, who took in weaving and finishing work on a freelance basis. These people were the first to be affected by the mechanisation of spinning and weaving. No longer were they asked to take in work; they were now expected to work in a mill and become one of a dozen, perhaps even a hundred, workers.

Within a few years of the first steam-powered mill operating off Thornton Road, Bradford became home to dozens of such establishments, all specialising in the woollen and worsted fabrics for which the town was already famous. Before the Industrial Revolution, Bradford was a similar size to other West Riding towns such as Halifax and Leeds, which also produced textiles. But it outgrew them all: Bradford's uniqueness came from its geographical position, the willingness of the town's business-men to participate in worldwide trade and the speed at which the town could absorb immigrant mill workers.

Plush velvets, soft tweeds and feather-light silks were just some of the materials made into dresses, suits and saris for people of all nations. But there was one fabric for which Bradford became particularly well known: worsteds, which is similar to woollen but with combed fibres, giving it a smoother appearance. Worsteds were used for suits, dresses and uphol-stery. The city also developed a good reputation as a quality dyer of the fabrics it produced, spawning a whole new industry and providing many more jobs.

By the 1850s, there were already over a hundred mills in the town centre. The second half of the nineteenth century saw several Bradford businessmen accumulate vast fortunes through the textile trade. Men such as Samuel Cunliffe Lister, Titus Salt and Jacob Behrens were business barons in the Victorian age and made enough money through the wool trade to retire several times over.

The great enterprises they had created and handed on to their descendants were still in existence when our story begins in the 1920s. But what about the thousands of people who worked in the mills? What was life like for them? What happened during a working day? How did they spend their free time? The answers to those questions can be found in the pages of this book.

Every area of Bradford is covered but some mills are better represented than others. Larger enterprises such as Salt's and Lister's had thousands of employees and this is reflected in the number of people who came forward to be interviewed. Nevertheless, there are also lots of fascinating stories from smaller concerns and all of the stories have a common thread: working in a mill was hard work but it had its compensations; not in the wages, but in the comradeship. No one held back in describing what were – to modern eyes – difficult and sometimes dangerous working conditions. But the tales, both humorous and sad, tell of a world that has gone for good.

Bradford now has only around a dozen working textile mills. For anyone who did not live in the period from the 1920s to the 1970s, it is difficult to imagine what the city was like. At one time anyone standing on a hillside overlooking Bradford – perhaps enjoying the spectacular vista from Undercliffe cemetery – would have seen a forest of mill chimneys rising from what seemed to be every other street. Each of these chimneys was belching fumes out into the atmosphere, creating smog that at times made it hard to see your hand in front of your face. The streets literally echoed to the sound of looms running to produce thousands of yards of cloth every day.

Most people working in a mill were unaware of the finer workings of the textile trade and the people who ran it. They did their jobs as well as they could and received a wage at the end of the week. They accepted that the industry could be fickle and that, at times, they could be laid off. Then, just a few weeks later, they would come back to do more overtime than they could wish for.

The neighbourhoods around a mill were close knit in a way that is unimaginable today. On most streets, the majority of people were employed directly or indirectly by a mill. If there was a boom or a bust, the entire community would feel the effects. Even local shops noticed the difference, because of the effect on customers.

Children were familiar with mill life from an early age. They delivered messages; they heard talk from the mill floor at the dinner table; they would play in the mill yard in the evening; and they were knowledgeable about textile terms.

The twentieth century saw some improvements in the way that workers were treated. But, as we shall learn, there were many instances of less than ideal conditions throughout our period. It is interesting that those who spoke about the 1920s and 1930s had often followed friends and relatives into a textile job, often at the same mill. But, as time went on, this was less common. There was a feeling that the mill was no longer a desirable place to work – especially after the second world war – and children were encouraged to set their sights higher.

From our twenty-first-century point of view, it is difficult to grasp the concept of how few rights workers had, and indeed expected. We are now used to laws and regulations that are specifically designed to protect us, whatever level we work at. Most mill workers of the last century didn't expect any of this, and didn't get it. There were strikes and disputes of course, but most rules were there to benefit the employer, not the employee. In an age when people could be dismissed on a whim, few could afford to stand up and be counted.

# 1

# the JOB

As well as woollen and worsted fabrics, there were many other textiles – such as velvets, silks and brocades – produced in and around the city of Bradford. Some mills, such as Lister's, were famous for the production of a particular fabric: Lister's velvets were famous throughout the world, and the mill even provided velvet curtains for the White House in Washington. Many mills concentrated on one process, such as weaving; others, like Salt's mill, had the entire textile production process under one roof.

This diversity meant that a bewildering array of skills were required. From taking delivery of a consignment of raw wool, to weaving the yarn, from checking the completed fabric for flaws, to being involved in the export of finished goods, there were literally dozens of different jobs and specialisations. Some required little skill, while others needed years of training and experience.

## Jobber lad

The most junior job in a mill was that of jobber lad. As the name suggests, this was a position for a young boy who had just left school and was at everyone's beck and call. As the oldest of three sons, Howard Rudge had a rude awakening on his first day as jobber lad at Drummond's mill:

> I suppose I thought I ruled the roost at home. Both my mum and dad worked and so I looked after my little brothers a lot and I was quite bossy. That's why I got such a shock on my first day at Drummond's.

My dad had got me the job. He worked at Ripley's dyehouse and knew one of the overlookers at Drummond's. This overlooker gave me the briefest of nods, put a sweeping brush in my hand and said 'get to work, lad'. After I'd swept the length and breadth of the shed, trying not to get under anyone's feet, I was glad to hear the shout of 'break time'. I straightened up, expecting a welcome cup of tea, only to find out it was going to be me making it. Now, brewing up in Bradford must be one of the worst jobs going. These workers were so precious about their cuppas, there were so many different ways to make the drink and each person expected their treasured breaktime brew to be just right.

There I was with a huge urn, wanting to cry as everyone gathered around me, pushing and shoving, wanting to be first with their tin mugs. Eventually a tough-looking lady took pity on me and got everyone to queue up. It took me so long to serve all the drinks that mine was clap cold by the time I got it.

Howard had taken the jobber-lad position hoping to learn something about the textile trade, but was to be disappointed:

It really was skivvy's work. If I wasn't cutting up hard, grimy soap for the toilets, I was chasing a rat down the cellars. If I wasn't shoving a great skip of bobbins into the lift, I was cleaning oil off the stairs, getting trodden on by a tide of people. Most jobber lads just did the work until they were old enough for an apprenticeship somewhere else and then cleared off.

Leslie Boocock started mill work in 1935 as the eldest child in a family of eight who lived in Clayton. He had to go out to work at the age of fourteen to bring money into the house and remembers that anyone who wanted work had to be recommended. His name was put forward as someone of good character by a neighbour who was a mill overseer.

As a junior, Leslie was always chosen for grimy or difficult jobs that older employees didn't want. The mill was whitewashed once a year and at whitewashing time the young lads had to cover all the machines ready for the sprayers to come and do the work. This was a dirty job because of the oil and grease that covered the machines and the lads were given an additional half a crown, known as 'muck money'. No one told Leslie about this extra pay, so when he got his wage he thought he had been overpaid and took it back to the wages office.

He remembers that the money he earned helped his family a lot but the walk to and from work in all weathers could be very hard. Outside work, he joined the Scouts and was also a member of the Clayton church choir. In 1938, Boy Scouts were being recruited into the Air Raid Protection messenger-cycle service, which Leslie joined, and he went for first-aid training and fire-watching duty.

## Weaving

In most mills weaving was the most labour-intensive task. Larger mills such as Drummond's and Lister's could have hundreds of looms going in one room, a process that created incredible noise. Weaving is the process of interlacing two sets of yarn – the warp and weft – on a loom to create cloth. The material that results from the weaving could be plain or patterned and the quality and intensity of the weave would vary depending on the requirements of the customer and the purpose for which the cloth was intended.

The best way to learn was to stand beside an experienced

weaver and picking up the necessary skills by trial and error. This is how Jean Squires got into the textile trade. Jean left school at fourteen in 1941 and went to work alongside her aunt, who was a weaver. They worked in a room where two men did the twisting. Jean's first job was as a reacher.

She used to ache from sitting on a wooden stool all day while she wasn't needed. When she wasn't sitting, she had to clean the shafts of the fluff and down as they were taken from the looms, sweep the weaving-shed floor and take the rubbish outside and burn it. On a Wednesday her job was to go to the post office and get the insurance stamps. Another important task was to go to the baker's and the fish-and-chip shop to get the dinner for the burlers and menders, the weavers, winders and people in the finishing room.

Diana Roberts remembers numerous arguments with those in charge of the weaving shed in the mills she worked at about how many ends a weaver should be looking after. The number of ends would depend on the type of material being produced by the loom. Experienced weavers could be watching dozens of ends, ready to replace the bobbins when they became full. But arguments could start, said Diana, when people were off sick or on a break:

> Some of the overlookers would expect you to take charge of your own loom and other people's as well. If someone was off sick, they'd just say casually, 'you can run so and so's loom, can't you?' And there'd be no question of extra pay, you were expected to do it alongside your normal job. Sometimes you'd be run ragged, dashing about between two machines, supervising someone learning the job and trying to make sure everything was running smoothly. These arguments over ends were a thing that would come up with the unions time and time again but

I never remember anything being resolved in my time in the trade.

Molly Carter found that weavers stuck together and it was often difficult to get to know people from other parts of the mill:

As soon as I started in weaving, it was just that: weaving all day and never seeing the other parts of the process. The odd time I might go to another department on a message and catch a glimpse of dyeing or spinning, but that was it. You never got to find out much about what other people did.

Even in the canteen all the weavers would sit together on the same few tables and not mix with anyone else. There were pubs in Bradford that a weaver wouldn't go into because they knew they would be full of dyers and they didn't want to mix. It sounds incredible when you say it out loud like that. But it really was that strong, you stuck with the people who you worked with.

Before Molly was allowed to begin weaving, she worked as a rubber picker. As she said of the job:

It was the lowest of the low as far as I was concerned. All you had to do was pick rubbers out of the back of the machine so there'd be no lumps in the wool and you had to keep on with that job until you'd got the hang of it before you could move on to anything else. It was noisy work, leaning right into the machine and so boring. But if you didn't do it properly, you'd get earache from the weaver on that loom because the wool would be coming through with lumps and they couldn't weave properly.

I remember one of the girls I was rubber picking with was about seven months' pregnant. She was finding it really hard to lean into the machine because of her bump. I told her to ask the overlooker if she could do something else because she was in real discomfort. 'If you can't do the job, lass, then there's plenty who can,' was his reply. In other words, she'd be out if she didn't carry on. I felt so bad I'd encouraged her to complain. I naively thought they'd sort something out for her but it wasn't the done thing in the 1940s.

Women worked until literally days before their baby was due, to earn as much money as possible before they had to finish up. They were expected to lift and carry just like everyone else, but in most cases there seems to have been a real comradeship with fellow workers. Many people would carry out heavier tasks on behalf of a pregnant worker and even let them have a sit down while covering for them.

Kath Clough was a weaver for most of her working life. Like many others, she followed her mother into the mill. She learnt simply by watching the other workers and picking up the skills as she went along. In her view weaving wasn't something you could learn straight away; you had to pick it up by experience and you wouldn't be let loose on a machine for a while after joining. As well as being a highly skilled occupation, weaving was very physical and workers had to be fit, with good eyesight. When people first joined a mill they would be tired out using muscles they had never used before and would ache for days.

Once Kath was fully trained she was looking after six looms at once. She remembers it helped if you had a good eye for when something was about to go wrong. Sometimes when you went to the toilet, a couple of the machines went wrong in just those few minutes. One of the most common problems was known as a 'smash': this was where the shuttle wouldn't go across the loom, but would stop in the middle and all the threads would break. When this happened the machine had to stop and the weaver lost money.

A 'smash hand' was a worker who was paid roughly the same as a weaver but was employed solely to sort out this type of breakdown which, like weaving, was a highly skilled job. The smash hand used a reed hook and a chart to put all the threads back into the right place before the machine was restarted.

The vast majority of weavers employed in a textile mill would work for 'piece rates'. This meant that as well as receiving their normal wage at the end of the week, they would be paid a form of commission depending upon their output.

The piece rate, which could vary greatly from mill to mill, was fixed by the employer and varied according to the complexity and density of the weave of cloth the employee had been working on. Therefore, the piece rate wouldn't simply depend upon the length of cloth produced, because someone working on a more complex weave would take longer to produce the same length of fabric.

The fact that piece rates varied meant that some workers would ask around to find out the best rates and move mills accordingly. Sometimes though, if someone was unhappy at a particular mill, as Hilary Simpson was, fear of the unknown could prevent them from moving on.

Hilary remembers feeling that she would have liked to leave her first mill, Holden's:

I'd heard that weavers were earning more at another mill further away from home and one of my friends encouraged me to apply there. I was all for it at first because I found Holden's a bit too big and wondered if I'd be happier in a smaller place. But in the end I was held back by fear. Firstly, I had a good group of friends at Holden's and felt that I might not be so lucky somewhere else. I mean, sometimes you'd see groups of weavers on the bus and they'd look really rough, like you'd never get on with them. I was quite a quiet person and knew I wouldn't stand up for myself.

Secondly, I didn't know if I could go through the ordeal of an interview at another mill and then having to tell my boss at Holden's I was leaving. It was fear of the unknown and it was what kept some people in the industry for years. Some people liked the mills because the employment was regular and also because of the fact that mill work meant fixed working hours. If you worked in a shop or office, you might have to keep on going until the work was done. At a mill, when the buzzer sounded, you were free to go. You were governed by all these rules and regulations and to some people like me, it was actually a comfort.

## Dyeing

The dyehouse, whether in a mill or on separate premises, was a male-dominated environment. By the 1920s, Bradford had gained an excellent reputation for the quality and colourfastness of the dyeing work carried out in the city.

Edward Ripley and Sons was one of Bradford's premier dyeing premises and was originally centred on Hustlergate. Ripley's had built its reputation by concentrating on dyeing garments black, the colour most favoured by the working class. Indeed such was the success it enjoyed that Ripley's dye works in the Bowling area of Bradford was once the largest in the world.

But many mills, like Lister's, had their own dyehouses on the premises, and were able to colour fabrics to the customer's specifications. Until the industry became mechanised and, even later, computerised, mixing dyes was a process that needed human supervision. This meant that dyers had to work with noxious chemicals on a daily basis.

To most mill workers, the dyehouse was out of bounds and visited only on rare occasions. Kathleen Shuttleworth remembers that she would occasionally visit Lister's dyehouse when making a charity collection at Christmas. She recalls that, before going in, she and a fellow worker would tie themselves together so that they wouldn't lose each other in the steam.

Ernie Blake remembers that the dye workers he encountered during his years at Illingworth's were particularly noted for their practical jokes:

You'd always hear about them playing jokes. People used to say that if you went into the dye house, someone would throw a jug of water right over your head and the steam would be that thick you wouldn't even be able to tell who'd done it. And there were tales of them pinning down new recruits, wrestling their underwear off them and dyeing it pink.

You'd see them walking home and they'd always stand out by the smell. It was a real strong chemical-smell, it must have stuck to their overalls. Their hands would be dark coloured with the different dyes and I bet they never got the dyes out of their nails. They tended to stick together, I suppose because it was quite specialised work and people were always complaining about the smell if they sat near them on breaks. I mean, what were the poor chaps supposed to do? No one could blame them for sticking with their own workers when people were like that.

## Warehouse work

Mills accepted and sent out deliveries as part of the daily routine. The raw materials needed to create cloth would be delivered and the finished product would be shipped to the customer. A large mill like Salt's would send goods right round the world; smaller concerns might deliver only to West Yorkshire. The same principle applied to goods coming in the way.

The location of Salt's mill was convenient for transport links. Its nineteenth-century founder, Titus Salt, chose the site because it was next to a railway line that connected to Bradford and beyond. It was also close to the river Aire and the Leeds–Liverpool canal, which both run within a few metres of the mill buildings. The mill might not have been in the city centre but it had transport links that were the envy of its competitors.

Derek Spencer worked as a despatcher at Salt's mill during the 1950s and enjoyed the feeling that he was part of a process that resulted in goods being exported all round the globe:

> I came to the mill from a boring job in a quarry and straight away felt I'd found my niche at Salt's. I loved looking at all the stamps and we had to mark the parcels with the name of the country they were going to. It was fascinating to think that the stuff I was parcelling up and sending off would be at the other side of the world in a matter of days.
>
> I had friends throughout the mill, so I knew about the creation of fabric from start to finish. And so it was nice to be at the end of the chain and know that the raw materials I had seen coming into the mill were now on their way out to the customers. We used to have a good laugh as well. Because it was a warehouse, we didn't really have as many rules and regulations as those who

worked in the mills. As long as the job got done, the four of us who did the work were left alone.

I remember the funniest thing was a trick we played on a young lad who came down on a message. He'd been sent from spinning, I think, to collect a box. We knew he was coming down and the lads decided to give him a scare. It was a huge box he was to collect and one of the chaps climbed inside. When this poor lad came to carry away the box, there was a big shout and out jumped this chap.

Well, the lad didn't know what to do: he screamed out with fright and nearly fell over. We were laughing our heads off. After a minute or two, when he'd got over the shock, he started laughing as well. And he took it in good part. I heard him laughing about it with his friends at dinner time. It was all just harmless stuff, but in a job like that it helped to pass the time and break up the day a bit.

## Woolsorting

Woolsorting was for the elite; it was more of a profession than a job. It was invariably the highest-paid manual work in a mill and required a high level of experience. The woolsorter's job was to select the different grades of wool from each fleece. The fleeces had been baled straight from the sheep and could be quite unpleasant in their raw state. There was also the risk of anthrax from the fibres.

The work had to be carried out in the best possible natural light, and also relied on a good sense of touch. Woolsorters could sort wool simply by how it felt in their hands. Douglas Downey worked as a woolsorter for many years and remembers that the work was paid on a piece rate and that the pay could fall dramatically during lean times.

Woolsorters weren't able to do the job in artificial light and so would work less hours – known as 'short time' – during the winter months. The sheds where Douglas worked had roof lighting and windows facing north, which helped to reduce the effect of sunlight and shadows. After the second world war – with the advent of electric lighting that could effectively replicate daylight – the woolsorters were able to work longer hours. Douglas Downey also remembers that nothing was ever wasted; there was always a market for wool products. Even the waste product left after the wool had been sorted – and known as 'shoddy' – went with any sheep droppings for manure. He found that, in the sorting trade, workers were either really busy with too much to do, or waiting around for something to happen. He recalls that the sorting sheds were a great place for debates and for broadening a person's outlook on life.

## Fabric inspector

Fabric-inspection work was the last task in the production process before goods were released to customers. Again, this was a skilled job and the type of work that could only be picked up by experience of the various textile processes and an in-depth knowledge of fabrics.

Steve Bowman is one of the few people interviewed who remains in the textile industry today and his trade is that of fabric inspector. After the decline of the textile industry in Bradford, he began to work on a freelance basis, allowing him to remain in employment. Steve's original training, which began when he left school in the 1960s, consisted of college courses where he learned about the technical construction of cloth, the different types of yarn available and the weave construction itself. He had to learn how yarns would behave and what they could, and couldn't, produce. This practical knowledge allowed him to spot faults in a piece of material and to determine their cause.

During his early years in the industry he would often travel to London to meet textile merchants at their places of work. These merchants were men who bought cloth from the mills where he was employed. It would be up to him to sort any problems that may have occurred with the finished textile product. When a fault was seen in the cloth, Steve had to be able to work out at what stage in the process the problem had occurred, in order to determine who was responsible for compensation. He particularly enjoyed the troubleshooting aspect of this work.

Steve explained that a piece of cloth could look perfect when it arrived at a dyehouse after being weaved. But when it was dyed, the material could come out with stripes on it that weren't present beforehand. He would have to explain to the buyer how this had happened and make sure the customer knew it wasn't the fault of the dyehouse. Sometimes he would have to go as far as sewing a piece of the faulty cloth and another sample piece of material and dyeing them. The fault would then show up again on the faulty piece but not on the sample, proving that it wasn't the dyeing process at fault.

## The gatehouse

Some of the most visible – and sometimes the most unpopular – members of mill staff would be those manning the mill gate. In Bradford, the mill gate was known as the 'penny oil'; in other words, the penny hole. This referred to the penny fine that was levied on latecomers.

Most mills were strict on timekeeping and a buzzer would sound five minutes before the workers were due at their posts. This could be heard streets away and anyone who lived nearby would have time to dash out of their house and into work. The person working on the gate was instructed to shut the gates as soon as the designated arrival time had passed. Latecomers would be sent home, in a practice known as being 'quartered'.

This meant that the unfortunate workers would effectively be locked out until after morning breaktime and lost a quarter of their wages as a result.

Every morning, the mill manager would come down to the gatehouse and tell the gateman about any vacancies for that day. People looking for jobs would present themselves at the gate at any time of day. Ronald Hoyle remembers an over-looker coming down to the gate after being told that a lady was looking for work. The overlooker only had one question: whether she was pregnant and, when she said she wasn't, the job was hers.

Roy Conway's grandfather worked as a gatehouse man at Salt's mill. He remembers his grandfather telling him that workers were summoned to work with a blast of a horn, which was actually a converted ship's siren and was worked by steam from the boiler house. Workers began at 7 a.m., when the gates were closed, and the first break of the day – which lasted twenty minutes – was at 8.30 a.m. Roy remembers that many women in Saltaire village rushed home during this break to see their children off to school. Family legend tells of his grandfather locking two of his own daughters out for being all of two minutes' late. They were quartered, with their father telling them they should get up in a morning.

## Behind the scenes

With many mills employing hundreds, and sometimes thou-sands, of workers there were dozens of people working behind the scenes to support those involved in the production process. Workers had most contact with the canteen staff.

Hilary Simpson's aunt worked at the canteen in Lister's for several years and she remembers getting extra rations, to the envy of those she worked with:

My auntie always used to say the canteen was the best job of all. She was very keen on cooking at home; you'd think she'd have had enough of it at work. But she was kind to me: she'd always put extra bacon in my buttie or give me the bun with the most icing on! It was a family perk.

The only thing she didn't like was when there was a big queue of people waiting to be served. You see, people only had so long to get their breakfast or dinner and they didn't take kindly to spending that precious time in a queue; they wanted to be sitting at the table eating. So they'd heckle and hassle the staff if there was a hold up. Believe me, there were some tough characters in there. But, interestingly, my auntie always said it was the women who made the most fuss and would be spiteful.

Sometimes she'd have to take afternoon tea through if the directors had visitors in and she'd love to set the tray out posh with all the nice china, instead of the rough stuff the rest of us had. When they'd finished they'd let the canteen staff finish off any buns or biscuits, another nice perk.

Cleaners usually started work after the mill had closed and mainly cleaned the offices, stairs and canteen areas. This was the type of casual job that could be carried out by women who weren't able to work a full day at the mill but who could commit to a couple of evenings a week. Andrea Ratcliffe's mum worked as a mill cleaner, on and off, for most of Andrea's childhood:

Sometimes, I'd go in with her while she did the cleaning. It was only for an hour or so when there was no one to look after me. To me it was fun; to her I'm sure it was hard work. I used to swan about, swinging-off chairs, sliding on banisters and she'd be there, working her hardest. But you don't see it as a child, do you?

Sometimes there'd be people working overtime in the offices and she'd get nervous about me being there with her, in case she got into trouble. Because you could be dismissed from that sort of job just like that. But I knew enough to keep quiet and even look like I was helping if I was forced to.

Whichever section of the mill people worked in, they would be working closely with their colleagues, often in a confined space and for hours on end. The next chapter will look at experiences of the working day.

# 2

# the working day

For anyone working on the factory floor at a mill, the working day was governed by the buzzer. There were clocks on the walls, but no one was allowed to finish for a break or the end of the day until the buzzer had sounded.

One theme that came through strongly from people who donated memories to this book is that mill work could often be monotonous but the jokes, laughter and sense of comradeship made up for what could sometimes be tedious work. Hard physical labour and less than ideal working conditions were made bearable by a good relationship with colleagues. There were obviously many arguments and, at times, fights, but a strong sense of more good times than bad.

## A shock to the system

Even if someone had been into the mill visiting parents from a young age, there was nothing as memorable as the first day at work. For most people, it was a real shock to the system and often quite frightening.

Joan Holmes vividly recalls her emotions:

I still haven't forgotten the terror of my first day. I couldn't eat a bite at breakfast, I was that scared. The overlooker met me at the door and barely spoke to me, he just set off walking down the room and I had to follow. 'You're with Nell' he said and walked away, leaving me with a woman who was turned away from me, working at her loom. 'Don't just stand there,' she said, 'come and give me a hand.' And that was my introduction to the place I'd work for the next seven years.

Few mills provided a formal training programme. Workers learnt the job from watching other, more experienced staff. The exception to this was working as an overlooker, which is covered in more detail in chapter five. Overlookers, who supervised groups of machines and their workers, were expected to learn the trade through a combination of practical experience and studying the technical aspects of the job at night school.

Lister's mill, as one of the largest textile employers in Bradford, had a training school on the premises. Here, workers could be taught the basics of the job in a few weeks and picked up the rest by following the lead of colleagues.

The fact that many weavers were on a piece rate meant that any distraction would inevitably reduce their earnings. A piece rate was worked out by the length of cloth produced at the end of a shift. The type of cloth and the density of the weave were also taken into consideration, as these would affect how long the job took. This is why people were often unhappy about having to train a new worker. They would be slowed down by the trainee asking questions and by having to do things slowly to demonstrate how the machines worked.

As an apprentice, working alongside an older overlooker, David Briggs found that some of the older women would 'mother' him and it was easy to settle in. He has fond memories of the first time he was given a present from someone outside his immediate circle of family and friends. A lady in her thirties, which seemed quite old to David at the time, befriended him. He was thrilled when she presented him with a navy woollen tie one Christmas and has kept the present since he was given it in the 1960s.

## Breaks

Break time was a welcome relief from several hours of hard work. Some mills started as early as 6 a.m. and the shift would run for a couple of hours before a break around 8.30 a.m. If there was no tea trolley, there was a gas boiler in each room so that people could make their own tea.

David Briggs remembers the tea trolleys coming round the mill at break times. Someone would usually spot one of the canteen ladies crossing the cobbles with the trolley and alert the others. The overlooker would bang on one of the steam pipes to tell everyone that the trolley was coming. There would then be a rush as people took up their mugs to get a drink of tea or coffee and also buy goodies like rock buns, dripping teacakes or biscuits. There wasn't an official place to sit, workers would just eat and drink around the machines when the trolley arrived.

In mills that didn't have a tea trolley, it was usually the responsibility of the young jobber lads to make pots of tea. Most people brought in their own tea mugs and washed them at the end of the day. In a big mill there could be as many as twenty younger workers mashing tea for the others.

Leslie Boocock remembers having a mashing tin, which was an oval-shaped, double-ended container made of brass that his mother filled with enough tea and sugar to make two cups of tea during the day. When he rejoined the textile industry after the second world war, he persuaded his firm to buy a gas cooker so that staff could warm meals they had brought in from home. Even as late as 1946, there was no canteen at his mill and workers ate meals by their machines or outside.

Lunch time was a more complicated affair as everyone was entitled to an hour's break. This had to be carefully managed so that the machines were still able to run. Arranging lunch cover was something that the overlooker had to sort out. If he had experienced people on the machines, he could trust them to look after two or more machines while the others were having lunch. Sometimes people from other departments would come in to help cover the break times and make sure that the looms were still running.

Some people took in the leftovers from their Sunday roast to have for lunch the next day. The Sunday roast was usually the most expensive meal of the week and so it would be expected to last for a few days. It could also be fried up into bubble and squeak or the chicken carcass used to make soup.

It was the job of the most junior workers in each department to take lunch orders and bring them back to the mill. For many, it was quite a lucrative trade, as Kenneth Nelson remembers:

> The rewards were so good sometimes we'd actually get into physical fights. Me and the other jobber lads would always want to go to the fish-and-chip shop to get the orders. It was a big mill and sometimes you'd be taking as many as 100 orders. You went to the chip shop that gave you the best commission. Obviously with you taking in such big orders, they were vying for your trade. Some paid you in fish and chips but I always went to the places that gave you money.

A sandwich round could also bring its rewards if the mill was near to a Co-op store. When items were paid for, a family's divvy number could be used, resulting in a bigger quarterly payment from the Co-op as a result of the extra orders.

David Briggs took his own cutlery to the canteen at the mill where he worked, as did many of his colleagues, simply to get served quicker. Because of the large number of people using the canteen, the turnaround on washing up the cutlery didn't always keep pace with the number of people queuing. If you had your own cutlery, you could simply get served and then go and sit down. The food at the canteen was all cooked on the premises and in the 1960s David could get a hot, three-course meal for 1s. 6d., the cost being subsidised by the mill.

Joan Holmes had one of her most embarrassing moments in the canteen when she'd worked at the mill for just a few months:

> I'd got my dinner and carried my tray towards the tables. I held the tray with one hand to get some cutlery and somehow lost my grip. It crashed to the floor and the plate with my meal on smashed. Everybody started clapping and cheering. There I was, standing in my long smock, with the long white socks I'd worn at school, not knowing what on earth to do. No one got up to help me so eventually I went to one of the canteen ladies and asked for a cloth. Luckily she came over to help me and by then everyone had stopped looking.

The canteen at Lister's started serving lunch from quarter past eleven and workers would come through in groups. Departments were split into two groups so that machines could be kept running.

In fine weather people would eat their lunch outside, sometimes sitting on the skips. Ernie Blake remembers that the groups could be quite cliquey, especially when you first started work:

> Me and another lad were standing about bored, so we decided to go and join a group of other lads a bit older than us. They were laughing and flirting with some of the girls from the weaving shed. We really wanted to join in but as soon as walked up, one of them said 'go away, kids' and all the girls laughed. We slunk away and didn't try it again.

Molly Carter naively thought when she started work at the age of fourteen that the toilets were where you went when you were answering the call of nature. How wrong she was. After a couple of trips, she soon discovered that they were a meeting place, somewhere to touch-up hair and make-up, to complain about people and to pass on, or catch-up on, gossip.

At first, I just used to wash my hands and go back to my machine. I wanted to join in, but I didn't know any of the folk they were talking about. But after a while, I got into the gang and I'd chat away with the rest of them. It was the first place I got to try on lipstick and mascara. I used to get dolled up and thought I was the bee's knees. Then at the end of the day I'd wash it all off so my dad wouldn't see when I got home.

Mill bathrooms were usually less than luxurious. Some workers even brought in their own supplies to make things more pleasant. Dee Rogers remembers that, when she was given a soap set for Christmas, she brought in a lavender bar because she was so fed up of the awful hard soap supplied by the mill. Employees also had to bring their own 'mill cloth', because there wasn't a towel. The sink was a long, trough sink with taps along its length. If the mill's soap was used, it was handed out in blocks and the jobber lad brought it round when it was cut.

'When the girls were going out after work,' remembers Evelyn Pearson, 'there'd be a rush for the toilets with everyone crowding round the mirrors. Some of the modest ones would change in the cubicles, but some would have no fear about stripping off in front of everyone and getting changed into the clothes they'd brought.' She remembers the overlookers were always on the lookout for people loitering or smoking in the toilets. Because they couldn't go into the ladies toilet, they had to shout through the toilet door to ask everyone to come out and, if they didn't, he'd have to send in one of the girls.

## Easily distracted

Another strong memory was that any distraction to the job in hand was welcome, especially for those not on piece rates. Many mills had frosted glass in the lower part of the windows so people weren't tempted to keep looking out, but they would often climb on to the window ledges anyway.

David Briggs remembers the wide window ledges that workers used to sit on. These were also used to store belongings, such as hairbrushes and make-up. There were hot pipes running the length of the room and, on a wet day, people would put their wellies and socks there to dry, giving the place the smell of a launderette.

Another ruse to stave-off boredom was to pretend you needed to go get something from another department, according to Joan Holmes:

> If ever the overlooker asked someone to go on a message, we nearly fought over it. You'd always be up for going for a walk so you could have a word with friends in other departments or wander past lads you fancied. But sometimes it could be quite intimidating walking through some rooms, especially places like woolsorting or dyeing, which were only men. They'd shout out things to you or make comments on what you were wearing. Some of the language wasn't for the faint hearted!

If someone brought in a magazine, it would be passed around for everyone to look at. For those who were good at weaving, it was possible to read and keep an eye on the loom at the same time, although this wasn't allowed. Kim Andrews remembers taking it in turns with her co-workers to read or knit while the others kept an eye out for the overlooker. They'd have a pre-arranged warning signal and, when he passed the machines, everyone would be concentrating hard on the job in hand.

If someone had a birthday, it was traditional to bring in sweets or cakes. And there were other opportunities for tasty snacks, as Kim points out:

> Another thing we'd look forward to was if there'd been a boardroom meeting. Ours was only a small mill with a good family atmosphere and, after the meeting, one of the secretaries would bring round the sandwiches and biscuits that were left over. We'd finish them all and tell each other we could never imagine leaving any food over if we ever had a meeting in our department.

Bonuses would often be paid out as an inducement for an important job to be finished on time. Evelyn Pearson remembers that, if a bonus looked likely, it was sometimes necessary to take extra measures:

> If people were chatting in the toilets, we'd go and get them out. Even if they weren't bothered about the extra money, most of us were. We'd tell them to come out and pull their weight or not to bother ever asking for a favour again.
>
> When I was younger, a few of us would sometimes get together to keep the bonuses for ourselves, rather than 'tipping up', which was what we called handing over our wages to our parents. We'd all agree to take the extra money and say nothing. Because if one girl told her mum, you could be sure it would be all over the street in no time and you'd have to hand over the extra.

Joan Holmes remembers that one of the women used to sell cosmetics from a catalogue and would bring in brochures every few weeks:

> Everyone wanted to buy from her. At first, she'd pass the catalogue round and we'd look at it while we were working and tell her what we wanted to order. And then when the orders arrived, there'd be another fuss as the things were given out and people gathered round to see what you'd bought. People would be trying out lipsticks and perfumes. Our overlooker cottoned on to it and told this woman she could only sell at lunch times. She didn't do nearly as well after that, half the time we were only buying because we were bored.

## Clubs and treats

Many of the mills were keen to give something back to workers in terms of leisure and social facilities. The trend for this had been set in the 1850s when mill owner Titus Salt created the model village that served Salt's mill at Saltaire. The idea of providing employees with the means to enjoy their leisure time during the Victorian age was a new one, but was still going strong eighty years' later in Salt's and other Bradford mills.

Salt's was known for its brass band, which was made up of self-taught musicians who worked at the mill. They toured the city, particularly around Christmas time, giving concerts. Societies and groups were favoured by employers as they fostered a team spirit between workers and helped to ensure that even leisure time was spent productively.

Sports were catered for, with many mills providing teams for football or cricket matches between rival mills. Lister's was particularly well known for the variety of sports clubs there – including a ladies football team as early as the 1920s – and it even had its own playing fields for practice and matches. Those who weren't athletic could get involved by cheering on their work team on a Saturday afternoon.

The summer fortnight when the whole factory closed was eagerly anticipated. Until the 1960s, most workers were not paid for this time off and so could find it a struggle. The answer for many was to create an informal holiday club, handing over money to the overlooker each week. The money would then be paid out just before the holidays, giving back cash that would otherwise have been spent. This type of arrangement could also work well as a Christmas club.

Joan Holmes remembers that parties and dances were always popular at the worsted mill where she worked, and would be discussed for weeks beforehand:

> There'd be so much fuss before what we called one of the 'do's'. We'd talk about what we were going to wear, who was going with who and who we thought was going to turn up from management. If the overlooker was nice, they'd sometimes let you finish early on the day so you could go and get ready. The party was usually straight after work.

She particularly remembers a dance held during the second world war to boost morale:

> A rumour started going round the factory floor that some airmen had been invited. We were wild with excitement. Some of the girls were literally squealing and jumping about. It makes me laugh to think about it now. You wouldn't recognise some of the people you worked with, they'd got so dolled up.
>
> Anyway, they'd decorated the canteen so nice, with

bunting and bits of fabric that had been sewn into rosettes. At first we all stood round listening to the music. But then the door opened and in walked this group of airmen, in uniform, who been bussed in from a barracks near Harrogate, I think. Well, they must have wondered what had hit them, because every girl in that place made a beeline for them!

## Practical jokers

Anyone employed in a mill would sooner or later become the victim of a practical joke. With large numbers working together in confined spaces, a good sense of humour was invaluable.

The most memorable practical jokes were played on those people who had just joined the mill. Particularly vulnerable were young lads who had started as apprentices. Even if they had been warned what to expect from older family members, there was every chance they'd be caught out within their first week.

Brian Fraser worked as an apprentice for one of the biggest textile firms in Bradford. He remembers being told to go and get a bubble for the spirit level:

I thought something was up, as my brothers had warned me about being sent on errands. But it was really awkward, because the man I'd been paired with didn't seem like the type to joke about. Everyone glared at me when I hesitated to go on this message, so I set off to the workshop. When I got there, I nervously asked the foreman for the bubble. 'It's not ready, lad,' he said. 'But you can go on an errand for me while I get it. Go back to your department and tell them I sent you for a long stand.' I can laugh about it now, but I went back in all seriousness and gave them the message. It was only after I'd stood around for fifteen minutes or so that I realised I'd been had.

Another mill tradition was the initiation ceremony for apprentices. David Briggs has clear memories of such an event. When he first joined the mill, he had been told that there was an initiation ceremony and he was warned to be careful when going downstairs to collect his first wage. Sure enough, when he went to the wages office, a group of lads were waiting, ready to cover him in red grease. David had a lucky escape but other unsuspecting newcomers were caught out and had to wash themselves off in the basins before going to collect their wages.

Another dreaded event, particularly for women, was the birthday bumps. This was something that Dee Rogers hated. The trouble was, she remembers, the more you said you disliked it, the more likely it would be made into a big event. On your birthday a group of workers would catch hold of you by your arms and hoist you up into the air. Then they would bump you down to the ground, according to the number of years you were celebrating. Everyone would stop and stare and then clap afterwards.

## Trips and perks

Although some mills had active social and welfare committees, most people arranged their own outings. Some mills would have an annual trip that would be paid for by the employers and anything else was up to the workers to organise. Someone would pin up a notice with the price of the proposed trip and details and, if enough people signed up, they hired a bus.

David Briggs remembers one of his colleagues arranging a day trip to Blackpool. The man collected a total of £4 10s. from people who wanted to go. David became suspicious when he saw the man with cigars and nice jewellery just before the trip. On the day of the outing the promised coach didn't arrive and everyone was left waiting in the meeting place. The man had stolen their money and was never seen again.

One of the oddest perks was for people who worked in the woolsorting trade. They were believed to have the softest hands in Bradford. While other textile workers could develop rough, callused hands from years of handling materials and machinery, woolsorters benefited from the lanolin in the wool. The lanolin – still a main ingredient in hand cream – impregnated itself into their skin, forming a protective barrier against chemicals and moisture and kept their hands as soft as a child's.

Offcuts of material were one of the main perks of working with textiles. Even the smallest ends of fabrics could be used for a variety of things, as Evelyn Pearson explains:

> Most overlookers would let you take home odd bits of material that had been damaged or were slightly imperfect. In fact, we used to say they were mean if they didn't. You wouldn't believe the number of uses we got out of fabrics; dolls clothes, pram covers, tablecloths. I even made a little fake fur coat for my baby daughter from a bit of fur fabric. I'd seen the Princess Elizabeth wearing one just like it and I was over the moon to make something like that for my little girl.

Evelyn Hindle remembers that most mills had a mill shop for employees but doesn't remember the prices being much cheaper than ordinary fabric shops. David Briggs, however, recalls two more unusual uses for some of the waste products. When the leathers on a machine split they had to be taken off and replaced. The leather found a new life as a shoe sole. Workers would put their clog against the leather, measure it to the right size and then cut it out and attach it to the shoe with nails. Another perk was what were known as 'clinkers'; sheep droppings that were sometimes found stuck to wool in its raw state. Gardeners loved them and would take them home and then dip a bag of them into a water butt to produce manure.

## Wages

Opinions vary about whether textile work was well paid. Of course, the wages would vary from job to job. Jobs that carried a risk, or involved working in a dangerous or toxic environment, such as a dyehouse, were comparatively well paid. And people working at supervisory or management level could be earning as much as five times more than those on the shop floor.

Veronica Bennett found mill wages poor and remembered that the work was often exhausting. She began life in a mill at the age of fifteen, in the 1930s, and worked during the war years making military uniforms. She remembers often working seven days a week to meet the demand for uniforms. Even though she often got overtime rates, she found the work still wasn't well paid, as the industry had to compete with firms from overseas. However, she felt proud to be working with what were considered to be the finest velvets in the world.

Veronica remembers that, at one time, her wages were paid in metal dockets instead of cash. These tokens could be exchanged for groceries, but only in certain shops. This practice lasted until textile unions put a stop to it.

Most mills, particularly in the first half of the twentieth century, didn't use wage envelopes. Instead, wages were paid in a variety of ways. Ronald Hoyle was given a small brass disc with a number on it, which he had to present to the mill cashier. He then received his wages in a little metal tin. The money was taken out and the tin put back into a skip to be used again.

Other employers handed out money from a wooden bowl, with the coins folded into any notes.

Friday was wages day and the vast majority of mills paid wages on a weekly basis. This was the day when people would leave work and head off to the pub or the fish-and-chip shop for a treat before spending the rest of their wages on bills and household items.

# 3

# heat, noise, dust and danger

The environment in a mill was unlike anything experienced today. Whether in a large or small mill the dominant sensation was the noise, which was often overwhelming. This was something remarked upon by everyone interviewed.

Frequently, the noise did not stop outside of working hours. Many mills operated round-the-clock, particularly during busy times. This meant that those living near the mill would still be able to hear the machinery at home.

People who work in factories now are issued with earplugs or earmuffs, but they were virtually unknown during our period. In fact, the damage to ears doesn't seem to have been recognised or acknowledged in most mills. If someone went deaf in later years because of exposure to noise nearly every day of their lives, it was simply regarded as one of those things.

Evelyn Pearson became an expert lip reader after just a few years of weaving work:

> People would sing rather than chat while we worked, because it was easier to hear a song than a conversation. Having said that, if you were saying something bad about someone, they could probably lip read what you were on about anyway, so there was no getting away with gossip about the other workers. We did talk about each other, anyone does in a place where there are dozens of workers, but it would be during the breaks or after the end of the day. Sometimes we'd have a chat while we were waiting for the tram after work, telling each other about so and so who'd gone off with the overlooker, that sort of thing.

Another danger posed by the constant noise was of not being able to hear when accidents occurred. Looms would have emergency-stop buttons, but someone had to be aware that there was a problem in order to stop the machine. For this reason, one of the first things that immigrant workers who didn't speak English well were taught was how to say 'stop'.

As we will learn in chapter five, overlookers were responsible for fixing machines when they failed. Strictly speaking, the machine should have been stopped while the necessary repairs were carried out. But those on a piece rate would not take kindly to having to stop work for any reason. If production was halted, money was lost from their pay packet. Joan Holmes remembers that a female colleague lost her fingertip by putting it in the belt when her loom was still going, instead of calling the overlooker to fix the machine.

Experienced weavers would often be as knowledgeable about their machine as the overlooker and would carry out their own repairs if they thought no one was looking. Julie Dillon worked alongside an older woman who had been at Holden's mill for nearly thirty years:

> It used to take so long for the overlooker to come along when you'd called him over. I'd get so cross, because you'd tell him your machine had stopped for whatever reason and then you could see him through the glass in his office chatting to someone or on his phone. And you were just stood there, not able to work with everyone earning money around you. It was maddening.
>
> Anyway, Ethel worked two machines down from me and she was renowned for knowing what was wrong with a loom. She had really arthritic hands from working on the yarns for so long, but she was nimble and quick

when it came to the job. If you could, you'd get her to fix the machine without even alerting the overlooker. If health and safety knew about it, you'd be in big trouble. Supposedly we weren't covered by insurance if something went wrong. But Ethel didn't care about things like that. As long as you gave her a couple of cigs at breaktime, she was happy to help.

## Heat and dust

Perhaps the noisiest and most uncomfortable section of the mill was the boiler house. This was a male-dominated environment and the heat generated by the boilers that powered the mill machinery was intense. Some mills were coke-fired or coal-fired while others were run by water power. Harry Fuller would often be taken to the boiler house during the early 1930s as a young boy, to help out his father on a Saturday morning:

> The men who worked in there were filthy. They all had black faces, dripping with sweat. And their clothes? Well, it looked like they had never changed them.
>
> Most of them wore flat caps covered in dust and dirt and, when they smiled, their teeth looked so white against everything else. I used to go in with dad on a Saturday to make sure there were no repairs needed over the weekend. At first, the place terrified me, but after a few weeks I loved going. I'd sit at the side and watch my dad talking or sometimes I'd wander off and watch deliveries coming in and out.

Kathleen Shuttleworth has strong memories of the sound of the looms dying off each working night at Lister's. All the looms stopped when the electricity was changing over from the corporation power to the mill's own electricity in the evening. There was silence as all the power went off and the place fell into darkness for a few minutes before the mill generator kicked in.

Hannah Newton worked at Salt's mill from the age of fourteen. She remembers having to work to help her family to pay for a house. She started work on a bank holiday Monday, despite her father saying he wouldn't work on a bank holiday for anyone. She was given a brush to sweep up the fluff that had come off the machines. She hated this task and didn't like cleaning up after anyone, so her mum said she didn't have to go back if she didn't want to. Nevertheless, she went back and started to enjoy the comradeship. She recalls that in summer it was very hot in the factory and all the windows were opened.

Molly Carter remembers humidifiers being used and water spurting to keep the yarn moist and easy to work with. There was oil everywhere, which would also get on people's clothes, and it could be treacherous underfoot with the water spraying about as well.

Unless you worked in the mill offices, said Dee Rogers, everyone came home filthy:

It wasn't from any particular thing, just from standing by your machines and the general factory environment. You just took it for granted that anywhere you sat would be dirty. I remember once picking up what I thought was a ball of fluff from a bench so I could sit down for two minutes when my machine was running well.

I've never been so terrified; a rat ran out from this fluff. I screamed and threw it onto the floor and it ran out under one of the looms. When people heard what had happened, there was a general uproar. Some people were taking advantage of the panic to scream and just act daft, but a lot of us were really scared. Eventually one of the overlookers was made to go around with a brush under the machines and try and ferret it out. But it was never found.

A rat catcher was employed to deal with these vermin, a long-standing problem in mills. Rodents liked the warm environment, and the fact that bits of food were sometimes left lying about; best of all, they could create cosy nests in pieces of fabric. Someone could put down a sandwich on the floor, come back to it a few minutes later and it had been taken by a rat. Workers would shout for the overlooker if they glimpsed a rat crossing the factory floor while they were working. The overlooker would call the rat catcher, who would put down a trap. The trap caught the rat live and it was then carried out to the mill dam and released.

Kath Clough remembers that physical injuries were common, but workers would soon get used to the environment. There would be fibres flying around in the air, which wouldn't even be noticed after a while. After first learning to weave, hands would be bruised and people would get calluses on their hands from working the machines. Kath says that despite regular cleaning of machinery, there could be dust and fluff that would fly out from the looms and catch on clothes. Hair had to be washed every night if it was to be kept clean.

## Accidents

Because most jobs involved physical work, accidents were commonplace. Most people interviewed had either heard about, or seen, a serious accident in the workplace. Less serious incidents seem to have been shrugged off as a fact of working life. Nowadays, we are used to health and safety regulations and to first-aid kits being readily available. Although some mills made an effort to keep staff safe, there was usually no one trained in first aid and little in the way of first-aid provisions.

One of the most common accidents was a belt flying off a loom and hitting someone. Also on a loom, something as small as a nut or bolt or even a bit of fluff falling down into the machine could cause a shuttle to dislodge. The shuttle would then fly through the air and cause serious harm to anyone in its path. Many people hit by a shuttle were carried away unconscious. This is why such emphasis was put on keeping the machines clean and on regular maintenance.

Colin Hardy remembered a terrible accident when he was just seventeen years' old and had gone to fetch a tray of tea for his fellow workers. On his return, he stood at the top of the stairs and saw one of the minders trying to clean a machine while it was still running. His hand got caught and four of his fingers off were ripped off. The accident made a big impression on Colin and the memory is still vivid.

David Briggs's quick thinking after an accident helped to save a life, and his employers rewarded him for it. A young female worker accidentally slashed her arm with a Stanley knife while cutting an empty bobbin. David approached the scene and saw a group of women panicking around her. Although he hadn't received any first-aid training, he took her to the first-aid room and stopped the bleeding.

The woman needed eighty-four stitches and Briggs was told that he probably saved her life by his prompt action. He was awarded a length of suit material, which he still has today, and a book token – although he was a little taken aback when his manager told him he had to use the token to buy a book on textiles!

Elizabeth Graham remembers that there was a first-aid box at the first mill where she worked, but it was never kept stocked-up and sometimes people didn't even know where it was.

There was an incident when one of the older workers slipped on some steps that had just been mopped. The woman screamed and some of the workers heard her and ran out to assist. Her leg was twisted under her body and she had cut her hand while grabbing onto the banister. Someone asked who the first-aid person was but, in the confusion, no-one knew. Some of the men carried the woman down the stairs and she passed out.

Eventually it was decided that an ambulance was needed. Elizabeth stood outside waiting for it to arrive and remembers that nearly every window was filled with people looking out to see what had happened. The ambulance men asked if someone could go with the injured woman in the ambulance and a manager replied that an older member of staff could go and report back on what happened.

The woman who accompanied the ambulance was gone for most of the day but came back late in the afternoon and went straight to the boss's office. It turned out that the injured woman had fractured a bone in her leg and was off work for a long time. People used to take round baskets of food and tell her all the gossip at work. One of the cleaners was blamed for the accident. It was said that she had used too much cleaning fluid in her bucket. As far as Elizabeth knows, there was no question of compensation. The job was kept open for the injured woman and that was all she expected.

Molly Carter remembers an accident where a woman was doing the twisting next to a winding frame. She caught her finger in the machine and part of it was badly torn. She was patched up by the person responsible for first aid but it was bleeding badly. The next day the girl's mother came in and went mad with the overlooker because, she said, the girl should have been taken to hospital and the injury was now likely to leave a scar.

One of the worst mill accidents of the twentieth century was the Zetland Mill disaster of 10 January 1924. The Zetland building on Canal Road was a three-storey mill, staffed mainly by female twisters and spinners. On the day of the accident, cracks had been spotted in the ceiling, but no one seems to have realised the gravity of the situation. At 9.30 a.m., the whole mill collapsed, killing four and injuring twenty-one.

The only female who died was a Mary Anne Edmonson, who had five children who also worked at the mill. Help was quick to arrive and there were stories of brave rescuers injuring their hands and arms as they struggled to free people from the rubble. The Salvation Army and local clergy were quickly on the scene to offer comfort to the shocked survivors. More than six hours after the collapse, workers were still trapped and many who had escaped told of their terror as the ceiling caved in above them. The general view was that the accident could have been a lot more serious and it was only the effective work of the rescuers that prevented more deaths and injuries.

Alongside such terrible incidents were more lighthearted episodes, which provided people with more of a laugh than a shock. Molly Carter remembers one of her colleagues carrying a basket full of bobbins down a set of hard, stone steps and talking and laughing as she went. She lost her footing and fell the full length of the stairs. There was a gasp from everyone watching as the woman lay still at the bottom of the stairs. As they ran towards her, she suddenly jumped up and started laughing hysterically, saying she'd never felt so embarrassed. The next day, she proudly showed off the bruises on her hip where she'd caught the edge of a step as she fell.

## Health and safety

Lister's was one of the pioneers in offering care to its workers and had its own surgery staffed by a qualified nurse. All young workers had an annual medical until they were eighteen and there was also a chiropodist and dentist that any member of staff could use.

Despite mills presenting a fairly high fire risk, with oil and fibres in close proximity, fire precautions at many mills were rudimentary. They usually consisted simply of a bucket of water and a bucket of sand at the end of each machine, mounted on wall brackets. Mills were inspected at regular intervals but the inspectors always informed managers when they were coming, giving time to top up the water buckets and to ensure the sand was clean. Other mills were more sophisticated and had sprinkler systems, which were triggered if the room filled with smoke.

Diana Roberts said that a false alarm was something that was always appreciated by staff. When the fire bell rang, they all had to file outside and wait by the mill wall. Fire-safety staff would then tour the building to make sure everywhere was safe before letting the employees back inside. This procedure could take up to half an hour, providing the chance for a gossip and a rest. She remembers that one of the men from the engine room lit up a cigarette while they were waiting to go back inside. He was embarrassed when one of the fire-safety women ran up and asked if he realised how stupid he was being when there was a risk of fire.

David Briggs had always wanted to ring a fire alarm and got his chance one day when he heard crackling and opened a set of double doors to see a fire in a section of the mill. He shut the doors quickly and smashed the glass on the fire alarm to activate the bell. A number of managers came running in and one of them told David off for smashing the alarm, thinking he had done it as a joke. But it turned out to be a real blaze that damaged part of the roof and was attended by six fire engines.

## Keeping warm

The first people into the mill in the morning faced the building at its coldest. Boiler men, maintenance staff and overlookers were among the first to arrive. Brian Fraser had a twenty-minute walk to Holden's mill and recalls that, in winter, by the time he had walked from home in the snow or rain his hands were so cold he could hardly feel them. Many workers would leave their coats on until they had warmed up and the heating had worked its way through the buildings. In winter, they were allowed a cup of tea before starting work.

The radiators that ran along the side of many mill sheds were popular places in the winter months and people would argue about who was going to stand next to the hot pipes. They would be used as places to hang coats, so they'd be warm by lunch time, and also to dry shoes or gloves. Some people even left a pot of soup on the pipes, which would have warmed right through by lunch time and smelt delicious.

The great blizzards of 1947 caused real problems with snow and ice and many mills ran short of coal, even those like Lister's which had emergency supplies. Workers were still expected to turn up for work and many walked miles through the snow when buses and trains failed to arrive.

The whole mill was dependent on the workers who staffed the boiler house. If something went wrong with the mill's power, none of the machines would run and no one would be able to work. Evelyn Pearson remembers that a power cut could be frightening. It only happened to her a couple of times, once when she was on the stairs:

> I was just coming back from taking a message to the offices and it was evening, getting towards hometime. I looked out of the window and remember thinking it

was going to snow; the sky had that heavy look you sometimes see.

Suddenly the whole place plunged into darkness. The machines stopped and I could hear the girls screaming in the weaving shed and the voice of the overlooker telling them not to be stupid. All I could see was the reflection of the street lights at the end of the stair windows. I told myself I'd better not panic or I'd fall down the stairs. I grabbed hold of the banister and inched my way down the stairs, trying to work out when I'd got back to the bottom. I managed to get down to the floor where I worked and then went into the weaving shed. I didn't dare try to pick my way between the machines, so I just held on to the wall till the power was restored.

# 4

## BOSSES

Many firms such as Drummond's and Lister's passed from father to son. The eldest son usually took over the running of the mill, either when his father died or when he handed over the business. Sons who followed in their father's footsteps would have been very familiar with the business through being exposed to the textile trade from a young age.

Keeping it in the family also applied to workers. These strong family traditions created a loyalty on the part of both employer and employee that is rarely found today. It was a matter of pride for people to say that they had worked for the same family for generations.

Whether workers saw the mill owner or senior managers on a daily basis, they always knew if their employers were treating them well: some mill owners went out of their way to ensure that their workers were happy; others were not so accommodating. Management could make the difference between a mill being a good or a bad place to work. This was a period before there was any formal training on how to deal with staff and little legislation to protect the rights of workers. How a member of staff was treated at work could often depend on the personality of the manager.

### Staff matters

Few of those interviewed had day-to-day dealings with those who owned the mills, but many had strong opinions about the type of people they were working for. The provision of leisure facilities would always make a mill owner popular, but smaller details – such as subsidised meals, Christmas parties and even gifts to celebrate landmark events – made a difference to how an employer was viewed by his staff.

The vast majority of people worked for a male mill owner. Women may have reached the boardroom, particularly later in the twentieth century, but, in reality, running a mill was a male preserve.

Andrea Ratcliffe remembers that the owner of the city-centre mill where she worked was a heavy drinker. His drinking was well known among the workers, but never discussed in his hearing:

> We actually liked him and felt sorry that he had a problem with alcohol. According to the cleaners, there was always a bottle of whisky in the drawer beside his desk.
>
> He'd sometimes go out for what he called a working lunch and come back the worse for wear. He'd fall up the stairs as he came back and mutter something about the floor being slippy underfoot. We'd be looking at each other, dying to laugh. Once he got into his office, everyone would erupt into giggles. And at the parties we had now and again, he was always the first to be up and dancing. He gave us a real good laugh, bless him.

One theme that comes through strongly is that smaller firms were better to work for. Jean Mortimer's family worked in mills all their lives and her mother and auntie worked in textiles from the age of fourteen until they were made redundant close to retirement age. Their place of work was at J. & H. Fisher's Union Mills in Idle. Jean remembers the family-run firm was an excellent place to work, with everyone on first-name terms.

Hilary Simpson recalls how she found out that the elderly manager of her mill had passed away. She arrived at work one day to find all of the machines closed down:

I asked what was going on and one of the overlookers said 'the old man's died'. I was genuinely upset. He'd been a good, kind boss, letting me take time off when my mam had been ill. Not all of them would do that. Anyway, after a day or so a rumour started going round that our boss had left every single worker some money in his will. I couldn't believe that it could be true, but the place was buzzing with excitement.

Eventually one of the directors came in and stood on the top of the steps, banging his tin cup with a spoon to attract attention. When everyone was listening he announced that each worker had been left a bequest, according to how long they'd been with the firm. Even people who'd just done a week or two's service would get something. Quite a few of us went out for a drink that night, and he was the toast of the pub!

Roy Conway also has good memories of the management, this time at Salt's mill. When his father, Tom Conway, retired from Salt's, there was a collection by the workers and management and he was asked what he would like as a parting gift. He requested a Hohmer Melodeon, a type of accordion, as his old one had seen better days. The presentation took place in the boardroom and was attended by Tom's workmates, managers and some of the directors. After the presentation one of the managers asked him to play a tune and he played 'Sweet Sixteen', one of his favourites.

Many of the gifts given out at landmark events were treasured for years. It was customary for workers to be given a present from their company at the age of twenty-one and also on retirement. Ernie Blake worked for Illingworth's and has fond memories of the day he was shown into the boardroom on his twenty-first birthday:

I'd never been into the offices. It was fascinating to look round, even though I tried not to stare.

One of the directors, a chap I'd spoken to a few times, was with me and he produced a key and opened this polished cupboard. Inside were at least a dozen items and he told me I could pick whichever one I wanted. I remember seeing a vase, a watch, a tankard and a hip flask. I couldn't ever imagine myself wearing a watch, I'd have worried about ruining it at work so I picked the tankard. They had it engraved for me at Fattorinis in Bradford and I've still got it.

Leslie Boocock remembers that every Good Friday, his employers arranged with the local church to let each department go to a service at a different time and they were still paid for a full day.

At important times such as Remembrance Day, the machines would stop for a few minutes as a mark of respect. Royal events, such as a coronation or jubilee, were often marked by parties, both in the streets and at workplaces. During the 1930s and 1940s, the exterior of Lister's – also known as Manningham Mills – would be lit by fairy lights on special occasions, giving a display that could be seen right across the city. These little touches made workers feel happy to be working at Lister's and they had the satisfaction of knowing that the company they worked for was synonymous with quality: 'When I was at Lister's', said Hilary Simpson, 'I always felt proud to tell people that was where I worked. Everyone knew that Lister's produced the best velvets in the world and that they were worn in too many countries to count. Out of all the Bradford mills, Manningham mills was the one that people would instantly know.'

At Lister's in 1934, the Manningham Mills welfare committee was established and was run by volunteers who met out of hours. It was intended as a forum where problems affecting any member of staff could be discussed. Not every problem could be satisfactorily resolved, but the very fact that there was a committee was a step forward from the days when mill owner Samuel Cunliffe Lister had been in conflict with his striking workers in 1890. The people on the welfare committee were elected by their colleagues and there was also an accident prevention committee, which tried to anticipate problems before they happened. How much influence these type of bodies had is debatable, but anything like them would have been unthinkable a couple of generations earlier.

Some mills made a conscious effort to help their staff cope with family matters. Jean Mortimer's mother had four children under five when their father died. Finding holiday cover for the children could have been a real problem, but Jean remembers J. & H. Fisher's paying for all employees' children to be taken by bus to a local school during the summer holidays. Here, they were able to play in a supervised environment, without their parents having to worry about them.

## Hands-on

In an age before the human resources department, management styles depended upon individual personalities. Some mill owners were content occasionally to walk around the mill floor, speaking to members of staff like visiting royalty, while others adopted a hands-on approach.

Joan Holmes remembers one of the managers at Illingworth's made a point of being available to staff:

> He'd often come around unannounced, but not in such a way you felt he was spying. He'd stop by your machine and ask how things were going, and if there were any problems. And he actually listened to your answers. It

was all quite formal, no first names or anything, but we really appreciated him for it.

Colin Hardy worked a night shift during part of his mill career. He says that night workers were often more relaxed when the rest of the staff had gone home, as long as the job got done. He remembers working on New Year's Eve night and looking out of the window to see the boss arriving. The man didn't make himself known to the workers, but just peered at them from behind the machines and left when he'd satisfied himself that all was well.

However content a workplace was generally, there were always times when disputes arose. Kathleen Shuttleworth remembers that one of the ways the workers would show their contempt for a boss would be to all rattle their tin and spoon loudly as that person walked past. She remembers, though, that the managers would act as if they couldn't hear anything.

If a manager was approachable, it was possible to ask for favours. David Briggs once wanted to apply for a mortgage and asked his manager to write him a letter of recommendation to the bank. The letter his boss wrote made out that David was on a better salary than he was, so that he would be allowed to borrow the amount of money that he needed.

## 'It must be wonderful to live in Ilkley'

Although a mill owner or manager may have come into contact with his staff on a daily basis, most did not live in the same area as their workers. Victoria Bennett remembers that the directors at Lister's mill seemed like royalty and acted like real gentlemen. The suburbs such as Heaton, Baildon and places further afield like Ilkley and Guiseley were popular areas where the managers would live alongside other professional people.

Managers may not have lived nearby, but it seems that there was always detailed knowledge about their personal circumstances. Whenever the boss got a new car, moved house or had a new addition to the family, the news would spread quickly around the mill.

Kenneth Nelson remembers that he always enjoyed hearing how the boss's family were doing:

> You almost felt they were your own family, even though it sounds strange looking back. If you heard the boss's son had got into university, or the daughter was getting married, well, you'd be pleased for them, not envious. That just didn't occur to you. They were the people who ensured you had a job and if they were doing well, then your job was secure as well, that's the way I looked at it.

Kim Andrews and her friends were fascinated by one of the bosses at Drummond's:

> Now I look back, I reckon we had a bit of a crush on him. We knew that he lived in Ilkley and had two children, as we'd see them coming in to visit their father sometimes after they'd been to the grammar school. Anyway, we persuaded one of the girls in the offices to find out the boss's address and me and two of the other girls decided to take a train trip to Ilkley one Saturday to have a look at the house for ourselves.
>
> We all thought of Ilkley as a holiday place, it was somewhere we'd been taken for a treat when we were younger. We didn't have a clue where this address was, but we stopped a passer-by and got directions. Unfortunately for us, the house only had a name and not a number and so we were forced to walk up and down

this street, looking for the house. I'd never seen such a long road with so many fancy houses. This place was looking right onto the moors and it was beautiful.

We just stood and gazed at it for a while. There were two cars parked outside and I even remember little details like a pair of stone dogs outside the front door. We were scared we'd get spotted and after a few minutes we went away and spent the rest of the time looking round the town, thinking that it must be wonderful to live in Ilkley.

Another way that people could come into contact with management was through working for them in a private capacity. Julie Dillon's mother was offered a job cleaning at the home of the man who was one of the managers at Lister's, where her father worked. The house was in Heaton, only a few minutes walk from the mill, but a world away in terms of the type of housing most workers lived in.

'It was a huge stone semi,' said Julie. 'Mam got the job through word of mouth, that sort of work was never advertised, it went to someone the family knew and could trust. It was ideal work for my mam as she could fit it in around us kids and school hours.'

Julie used to go to the house with her mum to help out sometimes, although she now recognises that she was probably more of a hindrance:

I'd look after the brass ornaments and shine them until they were beautiful. Sometimes I'd dance in front of a big mirror in the hall, while mam was cleaning the stairs and landing. I always wanted to explore the house, but mam would never let me out of her sight.

I remember once we were cleaning the room of the

boss's youngest daughter. It was a beautiful room with a doll's house and a dressing table with a frill underneath. Her window looked out over the gardens, where there was a fish pond. For ages afterwards I used to wish I lived in that house and daydream that I was their daughter.

The lady we were cleaning for was very kind and she'd sometimes give us clothes that her girls had grown out of. They were always superb quality, and a bit too fancy for the type of life I had, but I'd always wear them. Even if it meant playing out in a velvet dress with a bow on the front, I didn't care, I felt honoured to have nice things like that. None of the other kids asked how I came to have them. Whether their parents told them not to say anything, I don't know. But we all had hand-me-downs, mine were just higher class!

It was Molly Carter's mother's local fame as a superb cook that brought her an informal job working for a director of Northside mills:

My mother had worked at Northside from when she left school, right up until she had her first baby. She sometimes used to bake pies and cakes and bring them in to share if it was someone's birthday.

The boss remembered her cooking and when he was having a party, or if they were doing a presentation for buyers at the mill, he'd pay her to do some baking from home. She'd make batches of pasties and little iced cakes, each topped with a tiny sugar violet. When she'd done them, I'd help her carry them to the mill, on trays covered with tea towels. As we walked through the mill, people would pretend to grab the food off the trays and my mother would swat them away.

It wasn't unusual for members of the boss's family to come into the mill for various reasons. Kim Andrews remembers a woman coming through Drummond's in the 1960s:

> She was wearing a long fur coat and a lovely hat. Some of the lads spotted her and started to whistle. She looked like she didn't know what to do so I went over to her. It turned out she was looking for one of the directors in the management suite and had lost her way. She'd only been in once before.
>
> When I got back to the machines, I told the lads who she was and they were worried she was going to say something to the managers and they'd be in trouble. She came back the same way and said 'thank you, boys'. Me and my friends laughed because they'd all gone so red.

## Going up in the world

Ronald Hoyle worked in the wool trade for around forty years, ending up as mill manager at Whetley mills on Thornton Road. His career path is a good illustration of how it was possible for someone to rise through the ranks to one of the highest positions in a mill. Ronald started his career in 1937 at age fourteen, and his first work outfit was short trousers and bib-and-brace overalls.

After the war Ronald returned to Northside mills, which had closed during the hostilities. There was no training for any of the mill jobs; you learnt from others as you went along. Boys would be taken on as school leavers to do menial jobs in the mill such as sweeping-up and going on errands. Most of them only stayed until they were sixteen and then a new batch of younger boys came in to replace them. Girls could stay longer as there were other jobs they could move on to, such as spinning and weaving.

His first step on the career ladder came when he was appointed to the post of improver-overlooker. This was the title given to trainee overlookers who had some experience and could work mostly unsupervised. After a few years as overlooker, Ronald became the assistant spinning manager, moving on to spinning manager when the post became vacant. He was then asked if he would take the job of mill manager, which involved being in charge of eight hundred people.

The mill manager was sometimes known as the 'walking boss'. He had his own office but a lot of time was spent touring the mill 'looking for trouble', as Ronald puts it. He would always make the effort to find out someone's name before speaking to them. This was in part due to an incident which affected his own style of management and stayed with him all his life.

The incident took place in the early part of Ronald's mill career, when he worked alongside an overlooker who didn't use the proper names of workers. If he wanted to attract someone's attention, he just shouted 'oi'. One young lad didn't like this and told the overlooker that he had a name and wanted to be called by it. The lad was sacked on the spot by the overlooker for his cheek. As the lad walked away, the overlooker shouted 'oi'. The lad turned around and the overlooker said 'I've got a name lad. It's God,' meaning that he had the ultimate power.

This little incident illustrates the power of overlookers to hire and fire at will. Ronald says that no one would have dreamed of going above the overlooker to make a complaint if they were sacked. After seeing the way some workers had been treated, Ronald always made sure that when he was manager, if he wanted to speak to workers, he would ask an overlooker for the person's name before approaching them.

Although management had the power to hire and fire on the spot, the flip side of the coin was that workers could also walk out of a job without giving any notice and be taken on at another mill, usually without a reference. Ronald had direct experience of how easy it could be to secure alternative employment. One day he had a serious disagreement with a manager while he was working as an overlooker. Taking the view that his working relationship with the manager had irretrievably broken down, he informed the man that he was leaving and gave the week's notice that was required of a supervisor. He approached a nearby mill, asked if they were taking on overlookers and was offered a job to start in a week's time. But before he could work the period of notice, his manager asked him if he would stay and he decided that he would.

# 5

# the overlooker

Of all the jobs in a textile mill, that of overlooker was perhaps the most challenging. As the supervisor of a group of workers and their machines the overlooker (almost always a man) was often a go-between, someone who had to balance the conflicting interests of management and workers. If the workers were unhappy, production could be affected and the overlooker would have to answer to those above him. If the management were dissatisfied, it was the job of the overlooker to convey this to the workforce and ensure that improvements were made.

Despite the difficulties of the job, there were plenty of willing applicants. Becoming an overlooker involved a thorough training in all aspects of textile production. Much of the training was on the job, but the aspiring overlooker also had to attend evening classes in his own time. With training involving up to three evenings study a week, the job certainly encroached on free time.

The personality of an overlooker had a marked effect on the morale of his staff and on the atmosphere in the department. As in all walks of life, there were personality clashes and disagreements. This chapter will look at the role from the point of view of both the ordinary worker and the overlooker.

## Hiring and firing

The main task of the overlooker was to support his workers and to ensure that the department ran smoothly. Most would join a mill as an apprentice overlooker, working alongside the existing overlooker to observe how he carried out his duties.

David Briggs remembers that lads who joined a mill straight from school took one of two career paths: either as the apprentice to a mechanic, or as an apprentice overlooker. Briggs chose the overlooker route because he felt he would have more contact with women, rather than being in the male-dominated environment of the mechanics' workshop. He remembers he fell in and out of love many times during his years in the mills.

Once he was fully trained, Briggs was expected to arrive at work before his staff. His first job was to go to the oil shed and pick up a can to oil the spindles. This was time consuming, as he had to oil 120 ends on both sides of twenty individual machines before the workers could use them. He also had to oil all the brasses on the looms.

His next task was to make sure that everyone had arrived at work on time and to sort out any labour shortfall. Briggs had his own small office at the end of the weaving shed, which had glass sides so that he could always see what was happening on the floor. He wore a short blue overall while the managers would dress in white overalls.

Briggs recalls that a messenger would arrive each morning from the manager's office. They would sort out between them whether there were enough workers to cover his department and whether any could be spared to help out in other departments that were short-staffed. Workers had to clock-in, so the overlooker could see at a glance if anyone was missing from the shift.

Harry Fuller was an overlooker at Holden's mill and remembers that, during his apprenticeship, he felt intimidated by the groups of women he was supposed to be supervising:

I was given a 'share' to look after. That was a group of twenty machines and I had to go around helping if the machines had broken down. If we couldn't fix the machine, then we had to send for a mechanic. But you were expected to try your best to sort things yourself.

A lot of these women were ten or twenty years older than me and they saw I was a bit shy. All the time they'd be making comments about how I looked and saying I was really sweet and so on. I didn't know how to deal with it. The overlooker that I was apprentice to, he was really bolshie and they were all quite scared of him, but no one was in awe of me at all! Still, they'd sometimes cover up for me if I made a mistake, so it wasn't all bad.

As an overlooker, David Briggs was also responsible for completing accident-report forms. The forms went into a lot of detail about who the injured person was, what hours they had been working, what shift they had been on and how far into the shift they were when the accident happened. He doesn't remember anyone claiming against an employer and believes that it would have been difficult to prove anyway.

The overlooker had the power to hire and fire and, for the most part, this was accepted without question. Although it was a serious matter for someone to be sacked, it wasn't perhaps as much of a problem as it might be nowadays. The sacked person could simply walk into another mill and be taken on without being asked for a reference.

Harry Fuller remembers that people would sometimes come back into work the day after they'd been sacked as if nothing had happened:

For the most part I didn't mind. It was usually just that I'd argued with someone and they'd answered back and I told them to get out. But if they did come back, usually I didn't mind. 'I thought I sacked you yesterday' I'd say. 'Well I'm back, I don't remember it,' they'd joke and everyone would laugh. It was worth taking them back on usually just to keep the peace.

Sometimes though, we'd take it a bit more seriously. If someone was always messing about and likely to cause an accident, they'd have to go. And there'd always be someone who'd stir up trouble and bad feeling among the other workers. Again, you'd try to get rid of them for the sake of a good working environment.

Machine maintenance was the key to keeping workers happy, as those who were on piecework needed the machine to be going so that they could earn as much as possible. If a pulley came off a loom, they would signal to the overlooker that they needed it replacing and he would put it back on while the machine was still running.

At lunch times, it was up to the overlooker to keep the machines going while making sure that everyone had a break. This, Harry Fuller believes, was where the more experienced workers were so valuable:

If you had people on your machine who had been doing the job for years, you were laughing. You knew you could trust them to watch a few machines and nothing would go wrong. But then there'd be times when you'd have quite a few new people and no one was up to keeping watch properly. You'd have bobbins getting full and no one to take them off, machines jamming up and when the person the machine belonged to came back from dinner, they'd go mad.

## Watch and learn

The quality of an overlooker's training depended on the overlooker they were placed with. Although an apprentice was expected to watch and learn, a conscientious mentor would make sure that they had plenty to do and stretch their skills. Kim Andrews remembers an overlooker on her shift at Drummond's who had no patience with any of his apprentices:

> He could be really sharp with them. These poor lads had come into the job straight from school and he'd expect them to pick things up in a flash. If they didn't catch on first time, he'd shout at them in front of everyone.
>
> We always felt sorry for the apprentices, they didn't dare answer him back and we'd take the lads aside and show them again what to do on the quiet, try to give them a bit more confidence. He was quite nice with us workers, but I think he thought he could get away with being a bit off with the lads because he knew they'd never complain.

To become an overlooker in Bradford it was necessary to be an apprentice member of the Bradford Power Looms Overlooker's Society and agree to go to Bradford Technical College. The college offered a variety of business-related courses to those willing to study in the evenings.

Particularly during the early part of the period, when skills in the textile trade were very much in demand, workers were very protective about their knowledge and who they taught. If an apprentice who wasn't a member of the Overlooker's Society was sent to watch a worker, that worker could refuse to show them anything until they'd joined the society. Sometimes people from outside Bradford who were taken on in a mill wouldn't be members and people refused to help them lift things or to take messages for them.

Overlookers who were still learning their trade could find they were earning less money than mill juniors. This was because the mill was paying for their college training. The evening classes were a big commitment over a period of several years and not everyone lasted the course, remembers David Briggs. He did a five-year apprenticeship at Bulmer and Lumb's mill between 1964 and 1969.

The first stage in his training was evening classes at college for twelve months, doing three nights a week. Monday night was English and liberal studies, Wednesday was maths and calculations and Friday night, textile, woollen and worsted. After the year was up, he was put on a full apprenticeship. He would do one full day at college on a day-release basis and one evening a week in class. The rest of the time was spent at the mill, learning the job he was training for.

Completing an overlooker apprenticeship did not guarantee the person a place at the mill where they'd trained. If there were no jobs at the time they had to look in the local papers, or rely on word of mouth, to secure a position.

Steve Bowman has spent much of his life in the textile industry. Although he still works in the trade, he has had to adapt his skills to meet new challenges. He has seen many firms come and go and watched mill after mill close down. He believes it is only his flexibility that has kept him in a job when so many others have been forced to leave.

His first job was at Shearbridge Mills on Great Horton Road. He left school in 1967 with a few O levels and felt that, at that time, school leavers were able to go into more or less any profession they wanted. He began an engineering apprenticeship but became disillusioned with the poor standard of the other apprentices at the night school. One evening he went home and saw two jobs in the paper. There was a trainee travel-agent position that paid £3 15s. per week and a trainee textile designer at £7. The attraction of getting over twice the pay of the travel-agent job was what made him apply and so he entered the textile industry.

As a trainee designer Bowman attended night school and worked on a day-release basis to get a City & Guilds certificate in textile design. The course at Bradford Technical College dealt with the production side of the industry. He ended up becoming a fabric inspector and worked at places that produced for the quality end of the market: in fashion fabrics, mohair, wools and baby bedding. Bowman became inspection manager at Denby's mill and believes he got the job because of his years of experience.

Steve Bowman had a team working for him but would also be involved in day-to-day inspection work, especially during busy periods. He had to make decisions on whether pieces of fabric were good enough for their end purpose. After seeing so many of the mills he knew close down due, he believes, to increased competition from abroad, he now works as a freelance textile consultant and again got this role by adapting his skills to suit the changing face of the industry.

## Good and bad

The people interviewed had both good and bad memories of their overlookers. One clear theme that emerged was that the job could be very difficult if the overlooker was unhelpful. But things were not one-sided; workers could make life uncomfortable for someone they felt was not doing a job well.

A good overlooker was regarded as someone who trusted people to get on with the job, but was there if needed. Hilary Simpson felt that one of her overlookers at Lister's was so good he could tell if a machine was about to break down before it did:

Sometimes he just knew by the noises that something was about to go. And so he'd sort it out before you got into work or after you'd gone home. Everyone wanted to work for that particular man because your piece rates would be sky high. The machines would hardly ever be switched off in his section.

A conscientious overlooker would be rewarded by having contented workers who turned up for work on time and were happy to take on overtime, as Joan Holmes confirms:

My overlooker, a man called Gordon, was one of the best you could wish for. He'd always make sure you had enough supplies of everything and that the machines were well oiled and there was no rubbish lying around to trip over. And if he thought you'd been working hard, he'd let you sit down for half an hour or so at the end of the day. Some of the others would get jealous, mind you, but he always had his eyes open and he knew who'd been working and who hadn't.

According to Andrea Ratcliffe, some of the worst overlookers to work for were those who'd only recently qualified:

You'd get some in who'd just got through their apprenticeship and started throwing their weight around. They'd shout at people for nothing, just to show their authority. And they'd never listen to advice about the weaving, not even from people who'd been doing the job twenty years. 'You're not paid to think' one of them would snap if someone tried to offer advice. But it nearly always rebounded on them and they'd have to come cap in hand once things started going wrong.

Most of them weren't bad; they just had big egos because they thought they had a bit of power.

We had a real laugh with the tricks we used to play on some of the new ones, especially the apprentice overlookers. I remember once we plaited a long tail out of some of the yarn we had lying about. One of the girls attached it to the overlooker's blue coat. He was walking around with it trailing behind him. He didn't know a thing till he went to the canteen and someone tipped him off.

Although there may have been laughs in the workplace, an overlooker who harassed or bullied staff was a different matter. Some had a reputation for being overly keen on touching female staff or on shouting at people who couldn't answer back. Molly Carter explains how this type was dealt with:

We had a system for alerting each other about this particular chap. It seemed like every time you were turned facing your loom he'd be sidling up, putting his arm around you and asking if everything was all right. So we used to give a sharp tap on the pipe if he was on his way. There was no point trying to alert each other by shouting, the machines were too loud. But you'd hear the bang on the pipes and by the time he got to you, you'd be standing facing him so he couldn't get up to his tricks.

Molly remembers one of the overlookers getting his beard caught fast in her loom when he was trying to see what was making a noise in the machine:

He leaned right round and the next minute he shouted

out loud. His long beard had got stuck. One of the workers stopped the machine and two lads had to come and help him get free.

He wasn't hurt, but he was really nasty about the whole thing. If he'd made a joke of it, we'd have felt sorry for him but he shouted at me and then stormed back to his office. Everyone was laughing and not even bothering to hide it, we couldn't believe how nasty he'd been.

Jean Mortimer worked with an overlooker who was disliked by most workers. She recalls the day that someone put laxative in his tea and he was hardly seen for the rest of the day.

In one of the mills where Evelyn Pearson worked, you had to put up your hand to attract the overlooker's attention if you wanted to go to the toilet:

Sometimes, they'd pretend not to see you. One chap would mutter that you should have gone during your break, as if you could control when you wanted to go. When you did get permission to go, the overlooker would turn a board at the end of the wall to 'engaged', as only one person at a time was allowed to be out of the room. I think they thought it would stop people chatting in the toilets.

Victoria Bennett remembers that the overlookers weren't allowed to swear at the workers or make threats. Some of them had a good sense of humour and would put requests across in a pleasant way. If they wanted to ask someone to work harder they'd say 'do you like working here?' This was a way of telling the person to watch their step.

Few people seem to have complained to anyone above

overlooker level about any aspect of their job. Because the overlooker could hire and fire at will, it was usually pointless to take a complaint any further. Also, an overlooker could make life very unpleasant for a person who had tried to complain about his conduct. 'They had ways of getting you out if they wanted to,' said Andrea Ratcliffe. 'They'd mutter things about you, never be available if you wanted help, even try to turn the other staff against you until you decided it was easier just to go.'

Julie Dillon remembers a particular incident when someone did seem to get the better of her overlooker. She thinks the events have stayed in her mind because they were so out of the ordinary:

> There was one girl who was being picked on by our overlooker; everyone noticed it. He kept telling her that she was fat and lazy and seemed to be singling her out for worse criticism than everyone else. One day she told us she was going to report him and went to the offices while he was on his break. When she didn't come back we were all wondering what had happened. We genuinely thought she'd been sacked.
>
> After a while she walked in to collect her bag and said she'd been given the rest of the day off paid. No one knew what had taken place in the office that afternoon, but you can imagine the rumours that were flying round. Anyway, the next day this girl came back as if nothing had happened and she said she couldn't tell anyone what had gone on. There was a rumour that she'd threatened to get the union involved but she always refused to tell anyone what had been said that day. The overlooker still said things to her like he did to everyone else, but he never singled her out again.

Whatever level mill workers were on, everyone looked forward to the end of the working week and leisure time. The following chapter will explore how Bradford's mill workers spent their valued time off.

# 6

# all work and no play . . .

Working in a mill was physically demanding, with long hours, but there were compensations. Everyone interviewed had good memories of the type and variety of leisure activities they took part in. Hardly any of these activities involved spending a lot of money or travelling a long way from home.

Bradford during this period had an excellent range of amenities for people of all ages. Even during the second world war, there was a good choice of things to do to leave the hardship behind. Nowadays, we are used to several weeks paid leave from work every year. But this is only a relatively recent development. Millworkers were only entitled to two weeks' holiday a year and, for most of the period covered in this book, that leave was unpaid.

Perhaps because leisure time was so scarce, people were determined to make the most of it. Community events such as fairs and parades were attended with enthusiasm and, as we shall see, weeks of preparation would go into organising such events. And the organisation of the big day was all part of the fun.

## Close to home

One of the most convenient ways of spending leisure time was at home. For people with young children, this was sometimes the only real option. But in a close-knit community, it would be easy to round up neighbours for card nights or suppers.

A night of cards was something that could involve the whole family, as well as friends and neighbours. The games included rummy, whist, solo and old maid. Someone would bring drinks and another guest would provide baked biscuits.

Dee Rogers remembers that, as the group of teenagers she hung around with grew too old to be playing out in the street, they would often go for a game of cards at a friend's house when they were still too young to get into the pubs. Her mum never minded this because she always knew where they were and what they were doing.

She recalls that a night of cards was the ideal opportunity to test the reactions of her family to a new boyfriend:

> I'd met this bloke at work, I must have only been sixteen or seventeen and I really liked him. He kept asking me to go out to the cinema with him but I felt too shy. I realised that if I could get him to come along to one of my mum and dad's card nights, I could relax and be myself without all the pressure of a big date.
>
> I was a bundle of nerves on the night. I'd done myself up as if I was going to the Ritz. My hair was all curled and I had a huge black-and-white skirt that I thought was so stylish. Anyway, in he walked and my dad offered him a beer. They got talking about football and it was fine from then on. As I watched him showing my little brother how to play rummy, I knew I'd be ok with him. At the end of the night he gave me a kiss on the cheek as I stood on the front step saying goodbye to him. My first thought was horror that one of the nosy neighbours might tell my mum, but I hoped it'd be the first of many dates. And now I watch him showing our grandchildren how to play rummy!

Another cheap, but friendly, night was a pot-luck supper. Kenneth Nelson remembers that these became very popular during the 1960s:

There was a group of us lived on Great Horton Road, all doing similar jobs with kids around the same age. One night, we were all out sitting on our doorsteps. It was one of those summer evenings when it stays light till about ten o'clock. One of the neighbours was moaning that she'd love to go to a restaurant but didn't have anyone to leave the children with.

Someone said, 'We'll have a pot-luck supper.' Well none of us knew what it was so he explained that there'd be five courses to this meal and each course would be served at a different house. Nothing fancy, just things that could be eaten easily with the children in tow.

Well, we arranged it and you know, it was one of the best things we ever did. The children thought it was wonderful, running from house to house, seeing what the next course would be. We started with prawn cocktail, which seemed very exotic in those days, and then one of the women did a really tasty sausage casserole. The last course was a surprise, marshmallows to toast on a little fire on the wasteland at the end of the street. I've been to some great restaurants but, in all my years, I don't think I had anything better than that.

Evelyn Pearson remembers that, towards the end of the 1960s, barbecues were becoming more popular, inspired in part, she thinks, by American films. If someone on the street was having a barbecue, everyone could smell it and would wish they could join in. There was little privacy on the row of terraced houses where she lived and it was common for neighbours' children to hang over the fence and ask for something to eat.

Bradford, like all other northern towns, had a good range of public houses. Many were named along a textile theme, such as The Noble Comb, Sir Titus Salt and the Hand and Shuttle.

Most of the back-to-back terraced streets had two or three pubs within easy walking distance and some were handily placed near to the mill entrances, ready to tempt workers inside. It was for this reason that many women took to waiting outside the mills on a Friday night to claim the wage packet before it was lost to local hostelries.

Pubs were sometimes a way for an unscrupulous mill worker to get rid of items that had been taken from work. Kenneth Nelson remembers a man regularly coming into a pub on Manchester Road with a brown-paper parcel under his arm:

> This bloke would go up to the landlady and she'd look at what was in the bag first. If she didn't want it, the bloke would then go round the tables offering whatever material was in the bag was at a cut price. If he sold it, he handed over some coins to the landlady.
>
> No one seemed to question how he got hold of this material or whether it was just a perk of the job. To them, it was cheap fabric for dressmaking or even curtains. Sometimes a few houses in a row could have the same curtains if there'd been a particularly good consignment of one fabric.

## Making music

Bradfordians have always had a reputation for enjoying music and for producing internationally known names in this field. As in other mill towns, such as Halifax and Huddersfield, brass bands were popular forms of entertainment, and they often had textile links. Many mills, like Salt's, had a band and for a lot of people, following the band to different venues to see friends or relatives perform was an entertainment in itself.

Pauline Chapman's mother was a member of the Bradford

brass band. She played a cornet and the band met regularly for rehearsals and for performances. Pauline's grandmother Ellen Akins left school at the age of twelve to work part time in the mill. She had a lovely singing voice and her workmates would ask her to sing requests while they were working. She didn't have much leisure time but when she'd married her husband Tom and had two girls, she would go to the Bradford City Band where her husband played, along with his brothers. Tom and his brothers were self-taught musicians and first played in a band at St Mary's school.

Pauline remembers her grandmother telling her about her grandad Tom playing the cornet and piano at dances. After her grandparents and their friends had finished at the local dance-hall, everyone would come back to their house and they'd pull back the carpet, play the piano and dance some more.

In summer, families would follow the band as it played in different venues. It went to many places in the local area including Sunny Vale at Hipperholme, Halifax and to Bradford's Peel Park. At Christmas the band would walk up Otley Road and play carols in the streets.

## Religion

Due to Bradford's large immigrant population, the nineteenth century saw more and more churches and chapels spring up around the city. By the 1920s, there were hundreds of places of worship around Bradford. Often, it was possible to work out the ethnic make-up of a particular area by the church, chapel, mosque or temple.

The first wave of church building by immigrants catered for the Roman Catholics who came to Bradford following the 1845 potato famine in Ireland. But the newcomers were not always welcome; the foundation ceremony of Bradford's St Patrick's church was marred by a protest from those opposed to the Irish settling in the city. By the 1920s, the situation had stabilised and for many people the church became a big part of their social life and not just somewhere to visit on a Sunday. Many people remember visiting church or chapel three times on a Sunday, and saw nothing unusual in this. Church was a big social event, with people dressing up in their Sunday best and greeting friends and relations on their way to and from the service.

Andrea Ratcliffe says that the first time she was ever invited out to tea was by the vicar and his wife, who she got to know from her Sunday school. The couple would regularly invite six or seven children at a time to have tea at the vicarage:

> I couldn't believe how good that first tea was. It seemed so different to how everything was at home. What I didn't realise then was that the vicar's wife didn't work, she was there to support the vicar and so she had time to make beautiful teas and have a lovely house.
>
> They'd set out the tea party so nicely, there was lovely china with a maroon flower pattern that I thought was wonderful. Slices of apple and pear – an unheard of luxury round our way and cakes that were so small they'd be gone in a bite. But I don't ever remember feeling nervous or out of place, they went out of their way to be friendly.

Church processions were a popular community event, and even involved those who didn't attend religious services. Every year the Roman Catholic churches would hold eagerly anticipated May processions. A statue of the Virgin Mary would be carried through the streets, followed by one lucky girl who had been picked to crown the statue at mass, and behind her boys and girls decked out in their best outfits. The girls wore white dresses that, if possible, were brand new. Hymns were sung and the procession would be watched from doorsteps, shop fronts and any other convenient vantage points.

First Holy Communion was another occasion for a white dress to be produced for Catholic girls. Boys would wear grey trousers, a white shirt and a red tie. Holy medals and prayer cards were often given out as souvenirs.

Church parade was held by Anglican churches and involved the Scouts, Cubs, Brownies and Guide groups of the parish. Hilary Simpson has fond memories of those special days:

> Our parade was once a month. I used to say to my mum that I didn't like the parade Sundays because the service went on for longer, but really it was good.
>
> We'd all be out wearing our Guide uniforms, eyeing up the Scouts. There was always a huge to-do about who got to carry the flag. One time, me and another girl were still pulling the flag back and forth between us when the parade got started. Our Guide leader went mad, said we were being disrespectful and, after that, we'd always have a draw to see whose turn it was.

Good Friday walks were particularly popular in Bradford's Catholic communities. There would be a mass in the morning and then the men and boys from the choir would take the train from Bradford to Saltaire. From there, they would ride the tram railway to Shipley Glen and walk across to Myddleton Lodge monastery for the three o'clock open-air service of Stations of the Cross. After this long service, they would set off home, stopping for an ice-cream on the way.

Ernie Blake has happy memories of his days in the Baptist church choir at Girlington:

To start with, me and a couple of mates joined for two reasons; firstly because we'd heard that the choir were taken on trips quite frequently and, secondly, we thought it would be a great way to meet girls. We went to an all-boys school and by the time I was fifteen, I thought I was ready for a girlfriend!

What we didn't realise was that the choir was quite a big commitment in terms of your leisure time. We'd have rehearsals a couple of times a week, you'd be singing at one or two services on Sundays and if there was a wedding, you'd be expected to perform at that as well. We did used to get paid for the weddings, just a couple of shillings each. I told my pal Tony that I was going to leave, it wasn't at all what I'd had in mind. But he wouldn't have it. He'd met this girl in the choir, Sylvia I think she was called, and no way was he going to the choir on his own.

After we'd been in the choir a few months, I was actually into quite a routine with it. I mean, what would I have been doing anyway during rehearsal time? Just sitting around street corners bored. I can't have minded going that much, because I joined the adult section when I turned sixteen.

Joan Holmes remembers that her church choir lost quite a few members when the second world war broke out and people were called up. Some of the older parishioners stepped forward to take their places. The church seems to have provided support and comfort for many Bradford people during the war years, with organised groups as well as the regular church services.

Fred Hanney had reason to be grateful to the members of the Mother's Union group at the Catholic church of Our Lady of Lourdes and St William on Ingleby Road. His wedding to

Mary Byrne took place in April 1940, when food rationing was already underway. The Mother's Union provided a slap-up wedding breakfast for the bridal couple, who had lived in the parish all their lives.

In all Christian churches and chapels, events like first communion, baptism, confirmation and marriage would be occasions that were a source of joy and involvement for the whole neighbourhood. Kim Andrews remembers that when a baby was born, it would be a cause for celebration involving the whole street:

> The christening would take place in the local church and all the neighbours would go along, even people who didn't belong to that church. They'd stand outside the railings and watch as the baby and its family went inside. And after the service they'd come back to the house with little gifts for the child. Sometimes I thought they were just going back for a nice lunch, but that's a bit unfair, people were genuinely happy and everyone brought a present, even if it was just a bib or a bonnet.

As we will see later in the chapter, Bradford had many fine cinemas and theatres. But some of the most eagerly anticipated theatrical performances took place in church halls. Weeks of preparation would go into these events, which were usually the retelling of a classic tale such as *Cinderella* or *Mother Goose* with lots of comedy thrown in. For just a few pence, people could see their friends and neighbours onstage, often wearing splendid handmade costumes.

Programmes were usually produced by members of the theatrical group (some of which have been donated to Bradford's archives) and they included the names not only of the performers, but also those who had helped back stage. Often there was also a photograph of the cast. The *Telegraph & Argus* was also invited along to take a photo at the final curtain call, which would appear in the paper a few days later.

## Sport

Over the years Bradford has produced many sporting greats, including boxer Richard Dunn, who has a sports centre in the city named after him. Sport during this period was very popular. As well as the various sports teams attached to mills like Salt's and Lister's, there were football, cricket and rugby clubs for all of the Bradford suburbs, with no shortage of fans.

Bradford's two biggest football clubs, Bradford City and Bradford Park Avenue, had been founded in 1903 and 1907 respectively, and had separate grounds. Before sport was disrupted in 1939 by the war both clubs would play in front of crowds of 25,000 or more.

Amateur sports teams may not have attracted such big crowds, but Roy Conway remembers his father was a big cricket fan and when Windhill played local side Saltaire in Robert's Park there would be around two thousand spectators.

The Bradford Dolphins water-polo team is one of the oldest in the country, founded in 1888. The club had an eager following from its earliest years. Spectators on the first few rows of the swimming pool would have to take along towels to protect themselves from being drenched as the match progressed. Fred Hanney remembers that the pool was covered with boards after the match and doubled up as a dance floor for evening events.

A less formal type of water sport was the unofficial swimming that took place in mill dams after hours, despite the fact that it was forbidden by most firms. Parents also warned their children about the dangers and Brian Fraser remembers his father would try to frighten him out of approaching the dam:

He told me that a huge snake lived in the dam at Whetley mills and, that when he was a boy, it had tried to pull him under the water. Well, that story terrified but fascinated me at the same time. So I'd often go along waiting for the snake to slither out of the water, but I never attempted to go in for a swim. Whetley mills actually allowed after-hours swimming and they even provided male-and-female changing rooms, but it wasn't for me after the tales my dad told.

As well as the weekly sporting fixtures, there were other events held at less regular intervals. Odsal stadium hosted sporting events, including showjumping, wrestling and the Asian game, Kabbadi. Edna Richmond remembers that the showjumping could be a day-long event and that many families took along a picnic to eat in the stands.

## Out on the town

As well as a night at the pub, Bradfordians had a good range of coffee bars, restaurants and nightclubs to choose from. Coffee bars were a popular option for those not old enough for pubs and clubs. Julie Dillon was a regular at the Olympus coffee bar on Great Horton Road:

> A whole gang of us used to meet up there. We didn't have much money; I was handing all my wages over to my mum and just getting spending money back. A frothy coffee had to last as long as possible, so you could stay in there as long as you could.
>
> We used to watch the older teenagers enviously and hear them chatting on about going over to the Mecca club later in the evening and about the acts that were

going to be on there. The waitress used to keep coming over to ask if we wanted another drink, but we'd always say we hadn't finished. To be fair though, they never hassled us, we stayed as long as we wanted. Even today it takes me ages to finish a cuppa – a legacy of those coffee-bar days!

Nowadays, Bradford has a range of international restaurants to rival any British city. With the various groups of immigrants, came cafés and restaurants offering cuisine from their native countries. The city now has more curry houses per head of population than any anywhere else in the country, but it wasn't until the 1960s that Asian restaurants began to open on a large scale. At first, they were used only by Asians but, as the immigrants integrated, word spread about how good the food was. Before long, there were over a hundred restaurants offering curries in the city.

Dancehalls were another popular night out. In the 1930s, venues such as the Majestic and Gaumont halls were well attended. During the 1950s and 60s, younger people began to join the dance scene, going along to dances to listen to their favourite rock-and-roll records.

## Park life

Many Bradford mill workers spent much of their childhood, did their courting and later brought their children up to spend time in local parks. Although the city still boasts dozens of well-kept parks, they were used on a much bigger scale in the decades before and after world war two.

Lister park is the most popular and best known and it also has a strong textile connection. The land for the park was sold to the city by Samuel Cunliffe Lister, of Lister's mill. Once completed, it boasted a large botanical garden, an open-air swimming pool, boating lake and even a weather-recording station.

William Edmondson has fond memories of the variety of activities the park offered. He particularly enjoyed boating on the Lister Park lake. There were various types of boats available to hire, as well as a motorboat that took fifteen passengers around the lake for 3*d*. each. He remembers that the owner and his assistant used to wear Royal Navy caps while operating the boat, which was called *The Princess Mary*.

On special occasions, such as a royal wedding, the park would be lit with fairy lights and, in winter, there was ice skating with firelit torches illuminating the skating surface. A chestnut seller would complete the idyllic winter scene. Even those who didn't own ice skates weren't put off, but would simply slide on to the ice in their ordinary shoes.

One of the best-used attractions in Lister Park during the summer months was the open-air swimming pool, or lido. Joan Holmes remembers the excitement of the pool opening for the summer:

> Every weekend we'd go down there and see if it was open yet. For some reason we never asked anyone in charge when it would open, we just used to turn up every week with our swimming costumes and towels under our arms.
>
> After the war, when it reopened, there were quite a few improvements and a diving board was added. It was a great place to see and be seen. Sometimes we wouldn't even go in the water, we'd just watch the lads showing off in front of us, doing dive bombs and pushing each other in.

For the less active, most of the parks in Bradford had a bandstand and concerts would take place a couple of times a week.

Those in Lister Park could attract thousands of spectators, some sitting on deckchairs, others just listening to the music as they walked past.

Even walking along the pathways of a park was an event, recalls Molly Carter:

> For teenagers, promming was the thing to do in the 1940s. It always started off as separate groups of girls and boys. We'd walk along in our best gear, eyeing up the other group but no one would dare to say anything. Then after we'd walked past each other a couple of times, one of the brave ones would shout something to the other group, and that's it, the ice was broken.
>
> We used to call it 'copping off' if you got chatting. Nothing like that phrase means today, it was all very innocent, just chatting and maybe a peck on the cheek. And then after a few weeks, the boy might ask you out on a date and you'd have to decide whether to tell your mam. You see, I never used to tell my mam I was going promming, she'd have gone mad. She thought I was just going for a walk. Nine times out of ten, I'd pretend I was out with friends even if I was going on a date just to avoid the questions.

Harry Fuller felt that the onus was all on the boys to make a move during promming, something which could be very intimidating when facing a group of confident Bradford girls:

> I might have seen a girl at work I liked and I knew she'd be going down to Peel Park; we all did at the weekends. And I'd plan it all in my head, what I was going to say, how I was going to be so cool and calm.
>
> But it never ever happened like that. You had to

Workers often had to stand at their machines for hours on end.

Happy young Bradford mill workers in the 1920s.

Workers leave Salt's mill after a long shift.

Many Bradford mills were decorated for royal occasions and special events, like this one in the 1920s.

Burling and mending was a specialised job,
requiring good lighting.

Weavers often worked in cramped and noisy conditions.

Woolsorting could be one of the more unpleasant jobs in a mill.

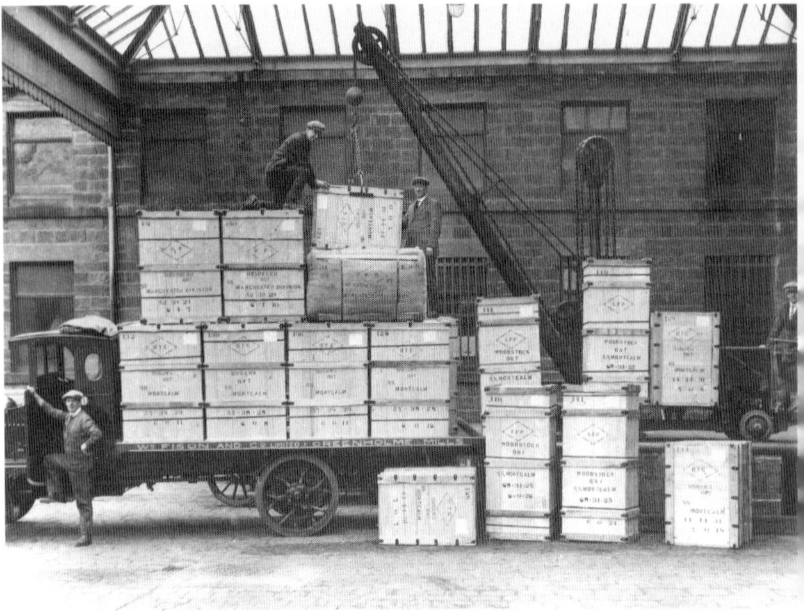


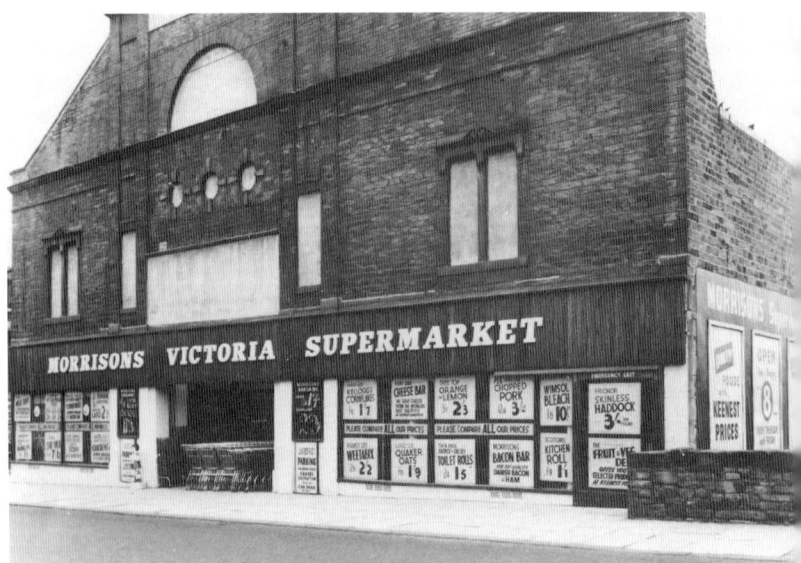

Many Bradford mills sent and received deliveries
from all round the world.

Morrisons Victoria store, the first experience of supermarket
shopping for many Bradfordians.
© Morrisons Supermarkets PLC

leave the safety of your own group of lads and walk over to the girls and try to separate your girl from the rest. Usually she would only talk to you for a minute, and then walk off with her mates and you'd be left standing there with your mates laughing at you.

It was the same carry on a few years later when we progressed to the nightclubs. You had to cross the floor to ask for a dance, the lady never approached and if she didn't want to dance with you, you had to walk back across the floor again and your friends had been watching the whole time. And to cap it all, when I did meet my wife, she told me she'd never liked nightclubs, she just used to go along so she wouldn't be left out.

## Cinema and theatre

There have been around thirty theatres in Bradford since the nineteenth century and none has been more famous, or more of a landmark, than the Alhambra. Completed in 1914, the theatre, with its three domes, quickly became a Bradford institution. Thousands of people had their first experience of being in the audience of a show or pantomime at the Alhambra.

Although Harry Fuller went to the cinema a couple of times a week during the 1930s, he found the theatre much more of a treat:

The Christmas pantomime was the best. Every child on the street whose parents could squeeze together enough money to take them would go. We always got the bus from Lidget Green in our best clothes. To pass the time, my mum would tell us to count how many Christmas trees we could see in windows before we got to town.

We always took our own sweets that we'd bought

beforehand at the corner shop. We weren't allowed to pick sweets with a wrapper as mum said they would make too much noise. I usually chose invalid butter toffee. I loved to watch the lady smash it with a little hammer and when you put a piece in your mouth, it would last for ages. Usually the actors would throw out sweets into the audience during the pantomime. Sometimes they'd even run out among the rows of seats and pick children to come up on stage and sing. I was always dying to get picked, but we were never near enough the front.

Every single time, my mum would say to my dad, 'I wish we had enough money to go in a box' and we'd all look up at the boxes at the side of the theatre. Dad said they could afford to have chocolates and drink champagne up there.

Kim Andrews remembers Drummond's mill paying for all employees and their families to go to the Alhambra pantomime one Christmas:

The kids were so excited. We had a big coach take us, we had to meet outside the mill and as soon as it pulled up, all the kids started screaming. When we got on, one of the managers handed each worker some money to buy ice creams at the interval, which I thought was kind. It was little things like that which made you appreciate where you worked, in comparison to other places you heard about.

The Alhambra hosted many big names, including Laurel and Hardy and Francis Laidler. During the fifties, the theatre also held talent contests to give bands and singers the chance to perform in front of an audience. There was a payment of a few pence to enter, and the first prize – decided by the loudest applause – was £25.

The thirties and forties were the peak years for cinema and there were forty-two cinemas in Bradford by 1939, many of them in the suburbs. The fact that the film would change twice a week shows how often people attended.

'There were double seats in the back rows at one of the cinemas in Girlington,' remembers Ernie Blake. 'Not that I ever got to sit in them. I always went to the cinema with a gang of my pals, very rarely with a girl. I liked to chat to girls and obviously you couldn't do that in the cinema.'

The period covered in this book spans the beginning of the talkies, through the war years – when films were popular as a form of escapism – to the films featuring stars such as Elvis Presley.

Mary Hanney and her brothers had a particularly memorable visit to her local cinema at Lidget Green in the mid 1920s. During the film, part of the roof collapsed and Mary's brother Laurence was injured. When the manager saw what had happened, he offered Laurence and his siblings free seats at the following Saturday's matinee. They enjoyed the performance and turned up again the following week, to be ushered in without paying. But when they turned up full of hope on the third week, the manager said he didn't remember them.

One of Bradford's most notable theatrical events took place outdoors, at Peel Park, in July 1931. The Bradford History Pageant involved 23,000 performers, including 7,500 school-children. Months of preparation went into the spectacular event, which involved displays of music and drama depicting moments in history from Roman times to the Industrial Revolution. The pageant was in conjunction with the Imperial Wool Industries Fair and helped to promote the city's wool trade by showing how it had become such a power in the world of textiles.

The final minutes of the pageant involved the children singing 'Land of Hope and Glory' by candlelight, and gradually walking out of sight. The sound of their singing growing fainter, and the light disappearing, must have been one of the most memorable and moving moments in Bradford's theatrical history.

# 7

# home life

Life at home obviously varied from family to family. But there is one striking fact: all of those interviewed said that people on the same street seemed to have a similar income. If times were hard everyone would struggle but, during times of economic strength, everyone would enjoy a higher standard of living.

Most mill workers lived in the back-to-back, or terraced, houses that still make up parts of Bradford today. These were the streets nearest to the mills where they and their families worked. Saltaire, the model village, is an extreme example of this pattern, but most people lived within a short walking distance of the mill where they worked.

Many people had an extra tie to work. Mills often owned whole streets of housing and rented them out to workers. Roy Conway's family rented a two-up-two-down house in Saltaire's Whitlam Street, which was named after Caroline Whitlam, the wife of Titus Salt. During the 1930s, the mill company offered their tenants the chance to buy their own house and Roy's family purchased theirs in 1937 for £295. The house had been modernised and had electricity, and there was a table-top bath in the kitchen and an outside flushing toilet. The house was painted and decorated inside and out and there was an Airedale-range coal fire to heat the water.

Some houses along Heaton Road belonged to nearby Manningham Mills. The advantage living in a mill house was that workers were assured of having a landlord that they could trust; their employer. The downside was that if workers were sacked, or laid off, they would be expected to leave the house with very little notice. Textile unions objected to these arrangements, as it made people heavily dependent on their bosses. But for those who wanted to live near to work there was very little choice.

Housing had been an issue of contention in Bradford ever since the Industrial Revolution, when thousands of new people arrived in the city every year. Bradford's population grew two and half times between 1801 and 1901 and much of the back-to-back and terraced housing was built to accommodate this huge influx. Most of these people had come to work in the mills, attracted by accounts of plentiful jobs and good wages.

By the 1920s, the housing situation had stabilised and earlier problems such as lack of sanitation and terrible overcrowding had been alleviated. Most homes still had outside toilets and many did not have an inside bathroom. The years between 1923 and 1937 saw a dramatic increase in house building with around 20,000 new houses being built to replace the slums.

Just as their ancestors had wanted to better themselves by leaving behind the slums, many workers longed to move away from the crowded streets around the mills and factories. Dee Rogers remembers her family visiting an aunt in Queensbury and returning to her house in Lidget Green feeling quite depressed:

It was just after I'd had my second baby and my aunt asked me to come up and see her. She said the air would do the children good. So I put the baby in the pram and I had a little seat for my son David, who was just a toddler, that fastened to the top of this coach-built pram. It was that heavy to push, goodness knows how I even made it to the end of the street.

Anyway, I managed to get a bus and I had to fold up this huge pram while holding the baby and keeping an eye out David didn't run into the road. Someone held the baby for me, and we got ourselves on somehow. I wasn't exactly sure where Queensbury was, but the driver put us off at the right stop and there was my aunt waiting for us. She took us back to this house, a semi-detached it was, and there seemed so much space. She had a back garden, a drive, garage, and the house overlooked open fields.

As we sat having tea, she talked about how much she was enjoying living there. They didn't have any kids, it was only her and my uncle. I knew that when I went home I was going to feel so bad because this place was beautiful. You could see right out for miles and it seemed so open. Walking down my street, still struggling with the pram and David grizzling away, I wished myself miles away! I can look back now and feel a bit sorry for her having no family except her husband but at the time I thought she had such a lucky life.

On a similar theme, Brian Nelson found that living on a street of terraced houses had its advantages, because everyone was in the same boat:

It was quite social really, no one could put on airs and graces because we all had the same as each other, two bedrooms upstairs, a flagged back yard and a front entrance with five steep steps leading up to it. You could sit out on your front step and look right down a row of identical houses, and every morning you woke up there'd be a row of the same houses staring back at you across the street.

And I tell you something, it was great for people watching. None of this sitting in cafés like they do today. As soon as a front door slammed shut, or you heard a shout, everyone would be at the window, craning their necks to see if anything interesting was going on.

## Outwork

Nowadays, many people dream of giving up their day job and working from home. For many families in a mill community, it was possible to secure casual, or outwork, from the mill. Because of the nature of the textile industry, dips in demand were inevitable. Mills could not always afford to keep on a full workforce and so work would be outsourced as required. Some women worked from home on piece rates, for example as out-menders, making repairs to fabrics that had been damaged during the production process.

Roy Conway's wife began her working life as a burler and mender at the Cooperative mills on Canal Road, just outside Forster Square. Sewing the broken ends into the finished pieces required a high level of skill and, for a while, she earned good money. When the Cooperative closed down, she and her friend worked burling and mending in what were called com-mission shops. These were dotted all over Bradford, generally run by a single owner, and were usually just a room over a shop. If the worker didn't like the owner, or if the piece rates were poor, they would start in the morning and not go back.

Once people had been trained to do a particular job, such as weaving, their skills could stay in demand, even when they had left to start a family or had retired. New mothers were kept on call on a casual basis to train new workers as and when required. They would work alongside the trainee until the per-son was capable of working alone and would leave again until more new people came in. Although they would still need to arrange childcare, it was easier to sort out for a few days than it would be for a full-time job.

A strike weaver came in on a temporary basis when someone was off sick. Again, this type of job was often done by married women who had left to have children and didn't want to come back permanently. The job provided extra income and children could be left with a neighbour because it wasn't on a permanent basis.

Older workers could also be retained. This was ideal for people who had retired but missed the comradeship. Ernie Blake felt that way about Illingworth's mill and was glad of the chance to come back:

I retired at sixty-five, went out on a high, I had a carriage clock presented, the works. The first morning, it was great to wake up and hear the buzzer going off, knowing I could just ignore it and stay cosy in bed. By day two, I was getting under the wife's feet and going mad wondering what they were all up to at the mill.

I missed the banter, the silly tricks we used to play, all of that. I got talking to one of the overlookers when he was on his way to the chippie and told him how I felt. 'There's still work for you, mate' he said. 'Not every day, but I could do with you coming in now and again. Can I give you a knock if I need you?' Well I was like a cat on hot bricks waiting for that! And after a few days a jobber lad came over to say they needed me because one of the looms had got jammed. I was over like a shot. After I'd finished, I stayed for a cuppa and when I'd had a bit of a chat with the other lads, I felt more myself again. Sometimes I wouldn't be needed for weeks at a time, but it just gave me a bit of a lift, knowing I could be called on.

Cleaning was another source of casual labour. Machinery running for around twelve hours a day and a hot, greasy environment caused a rapid build-up of grime. Even if workers were required to keep their machines clean on a daily basis, there was still the need for a good clean every few months. Workers would be asked if they had friends or relatives who would like to earn extra money for coming in to clean on the odd weekend.

The work was hard. It involved cleaning the mill floors using scrapers and caustic soda to get rid up of the build-up of grease. The same process was repeated in stairwells and cloakrooms. People with children would get them involved as well, with the younger ones passing equipment and older children involved in the cleaning itself. The alkaline solution used on the floors was harsh on the skin and many women would have work-reddened hands, from cleaning both at the mill and at home.

Another option for regular work in the home was taking in washing and ironing, as Diana Roberts did when her children were young:

My sister Jill was secretary to one of the directors at Lister's and she came home one night and said the man had been saying that his usual laundry lady was moving away and did Jill know anyone who could help? She thought of me, bless her, as she knew we were strapped for cash with me not working.

Every Thursday, a lad would drop off a parcel of shirts and collars. And I'd do them beautiful, the care I took with them, my husband used to joke I never went to that much bother with anything of his. There were five shirts each time and I used to think how nice it must be to change into freshly laundered things each day. I got three shillings a week, and it might not sound much, but it went a long way. It was mostly for little treats, chips from the chippy, the odd night at the cinema, just those little things that made the day-to-day routine bearable.

## Child care

For those not fortunate enough to be able to work from home, the goodwill of friends and family in looking after children was essential in an age before nurseries and childminders were common.

Evelyn Pearson left her first child with a neighbour when it was just a few weeks old. She remembers trying not to get too close to the baby as she knew it would be a wrench when she had to leave her and go back to work. She and her husband had explored every alternative to her going back to work but could not find a solution. So, when the baby was just six weeks, she left her with a neighbour.

Evelyn remembers handing the infant over wrapped in a shawl and then running down the street before she burst into tears. At breakfast she had to dash back to the house to breastfeed and then back again to do the same at lunch time. This left her with no spare time for a break and she had to wolf down her sandwiches as she dashed back along the road to the mill.

Another option was an early form of job share that was possible for people who lived near to each other. One worker would do the early shift at the mill while the other would look after that person's children alongside their own. Then they would swap in the afternoon. The system could fall down if one of the women was ill or if one of the workers was laid off, but it was a widely practised system that worked well and didn't involve childcare costs.

Older children were often a big help to working parents. Kenneth Nelson says that he practically brought up his two brothers and often collected them from school or a childminder. He made tea for the younger children and looked after them until his parents came home.

If children weren't old enough to go home and start the

tea, they would usually play out in the street or go along to the mill, remembers Elizabeth Graham:

> There were six or seven of us on my road who had mums or dads in the mill. After school, we'd be told to play outside, even in winter. We just hung around, swinging off lamp-posts, or made up imaginary games.
>
> If it was really cold, or raining, we were allowed to go and wait in the mill. But you were always under dire warnings not to touch anything or mess about. We'd usually just sit on the hot pipes at the side and let our coats dry out as we sat there.

In an age before maternity rights and benefits, women expecting a baby would work for as long as they could, to save as much money as possible before the baby arrived. If they wanted to go back to work after the baby, there was no automatic right to return. They would have to get someone they knew who could do the job, but wasn't actually working, to stand in and ensure the job was kept open for them.

Although women may not have been entitled to many benefits, Andrea Ratcliffe remembers the kindness people would show to people on their team who were expecting a baby:

> I remember my pal Hilary had really bad morning sickness when she was pregnant with her first child. She didn't dare to keep asking to go to the toilet because she thought she might be told to leave the job. So when she needed to go she'd tap one of the women and they'd alert the others that she'd gone. Everyone would keep an eye on her machine and try to make sure that the overlooker didn't notice she was away from her machine.
>
> It must have been so hard for her. She worked right

up till a couple of weeks before the baby was due. It was a really hot summer and the sweat would be pouring down her face. She must have been dying for a sit down, but there was no chance of that; she just had to keep on like the rest of us.

## Food

It was only towards the end of this period that supermarkets were established. The town centre had always offered a good range of shops, including John Street and Rawson markets, where fresh produce could be bought.

But most people who were working full time only had the chance to travel into the city centre on a Saturday afternoon, when work had finished for the week. For them, suburban shops were a lifeline. Many districts such as Lidget Green and Allerton offered parades with dozens of shops, selling everything from groceries to clothing.

In 1961, the first Morrison's supermarket opened in an old cinema at Girlington. This was the first self-service supermarket in Bradford, and it was the first of many around the city. Local merchant William Morrison built up his empire from a small stall in Bradford market in 1899. Many rival supermarkets appeared over the years, but Morrison's has always been regarded as a Bradford business by local people. The store also provided many jobs as the textile industry declined and people came out of their mill jobs.

For convenience, nothing was better than the corner shops that were found on most street corners in built-up areas. Although they could be expensive, they offered goods on credit to regular customers and opened long hours to suit the various shifts of the working people of the area.

Children were usually allowed to buy cigarettes for older

relatives at a corner shop if they were known to the owner and most children regularly went on errands, seeing it as an adventure. When children got home from school they were often very hungry, and a typical snack was a piece of bread and jam. Often they would run home from school, change into their playing-out clothes and then get a piece of bread and jam and run outside to eat it. They would be called in for their tea later in the evening when dad came home.

Hilary Simpson's father was keen on gambling and often tried to hide bets from her mother. He would send Hilary to the betting shop with instructions written on a note and the money folded inside. He promised that, if she didn't tell her mum, he would give her a penny for sweets.

Evelyn Pearson's trip to buy apricot jam turned into more of an adventure than her mother had intended:

I must have been six or seven, and mum was making a cake and asked me to go and get some apricot jam. It had to be no other jam, she made it quite clear. I got to the usual shop and they didn't have any, but they suggested I try a shop on the next street, which had run out as well. Instead of going home, like I should have done, I kept walking until I saw another shop. I walked right up to the start of Clayton village when I finally got this jam.

I felt so proud of myself, coming back with the right thing, but I didn't realise how long I'd been gone, you don't at that age. As I walked back into the street, I wondered why all the women were out on their doorsteps. Then someone shouted 'she's here, Marje,' and mum came tearing along the street and hugged me so hard it hurt. She was laughing and crying at the same time and asking where I'd been. 'You've been over half an hour, I thought you'd been run over,' she said. But then she smiled again and let me help her ice the cake, so I guessed I must have been forgiven.

Many children weren't allowed to talk during meal times, either at school or home. In Kenneth Nelson's house there weren't enough chairs to go round and so the younger ones would have to stand up to eat. There was a real emphasis on table manners and silence at the table. Sometimes his mother and father would talk, but the children wouldn't be expected to join in or venture an opinion.

Ernie Blake remembers being very proud when he came home from his first day at work. There was a special tea for him and he was given a piece of meat, just like his dad, while the rest of the family were having bubble and squeak. 'Mum always gave the biggest portion of food to my dad, but once I was working, I got just the same, and I felt so grown up.'

## Hand-me-downs

A common memory was the extent to which hand-me-downs were used for clothing children. As well as to siblings hand-me-downs could also be passed onto to cousins and neighbours; indeed to anyone who might fit into the clothes.

Many Bradford women were expert self-taught seamstresses, and a job in the mill came in handy for getting hold of fabrics. Some women would make extra money using fabric off-cuts from the factories that the overlooker had allowed them to take home. They would make pram covers, dolls' clothes, sunhats and anything that could be created out of a piece of fabric that was too small for regular garments. Molly Carter remembers her mum bringing home a really beautiful piece of ethnic fabric for her doll. The fabric had been woven with gold thread and speckled with tiny mirrors. Her mum showed her how to make the piece of fabric into a sari for the doll and Molly had the best-dressed toy on the street.

Another use for off-cuts of fabric was to use them for children's games. Pieces of fabric that had been stained or snagged were usually no use to the mill and so people would be allowed to take them home. Here, they could have all sorts of uses, including carpeting a tree house, making a den or putting in a dressing-up box.

Looking back, Diana Roberts realises she was lucky to have the handmade clothes that her mother worked on for her and her younger sister. But, at the time, she remembers feeling really embarrassed about it:

> Mam always made our dresses matching, in the same fabric, I suppose because it was easiest. But some of the girls at school would make fun of me and pretend they thought my little sister was my twin because we were dressed alike. One day I complained to mum and, from then on, she'd try to make the styles a little different, even though the dresses were from the same pattern, by changing the neckline or some of the decoration. Heaven knows how she didn't call me an ungrateful so-and-so but she was a very patient woman!

It was not only clothes designed for younger children that were passed on to siblings of the same sex. During particularly hard times clothes that had been made for the opposite sex, or even for adults, were reused in this way. Elizabeth Graham remembers her mum cutting up her wedding dress to make a first communion dress. Elizabeth says she didn't feel grateful at the time and was secretly annoyed that she couldn't go to the market and get a new communion dress like some of the girls in her class.

Only later did she realise what a sacrifice her mum made. Her wedding dress was the only quality clothing her mother had until the children had left home and they were a bit better off. She made the dress look so nice and Elizabeth now realises it was better than the shop-bought ones. Her mother sewed tiny fabric rosebuds from the draper's shop on to the hem and little pearl beads into the neckline. Later, the dress was used for dressing-up games and was particularly popular when girls played at being brides.

Whitsuntide was the one occasion when everyone who could possibly afford it would get a new outfit. Children would go around showing off the clothes to friends and relatives in the surrounding streets and would be rewarded with a penny or halfpenny at the houses they visited.

Shoes were another expense and even adults sometimes had footwear that didn't fit properly. Dee Rogers remembers putting sugar bags into her shoes because the soles had worn out and she thought the bags would be hardwearing. But the first time she went out in the rain the soles disintegrated, and she had to wait until she could afford to get them soled at the cobbler's.

Some children would be literally sewn in to their clothes for the winter. A layer of newspaper would be sewn between their jumper and vest to make one warm garment. Even though they crackled as they moved, the garment provided effective insulation for the winter months.

## Make do and mend

Some of the strongest memories of home life were special occasions. Several people remember that getting their first television set was a big event. Kenneth Nelson recalls that his family were the first on the street to get a television. But their initial viewing of the set was not to be a private, family affair:

Mum was mad about the royal family and she couldn't bear to think she wouldn't be able to watch Queen Elizabeth's coronation. She nagged my dad for weeks, we joined in too, and eventually he gave in. The TV arrived the day before the coronation and it was given pride of place in the living room. We kids gathered round eagerly, waiting for mum to switch it on. We couldn't believe it when she said no one was watching until the coronation next day.

Anyway, we went out to play and we were bragging to our mates, like kids do, that we'd got a TV. When it was time for the coronation next day, there was knock after knock at the door. In the end, there must have been about fifteen people crowded around that set. My mum had to make drinks and biscuits for them all and she wasn't best pleased.

For many people, having something new for the house was a rare treat. Some couples couldn't afford a house when they first got married and so would live with one set of parents. Even if they were lucky enough to get a house, often they wouldn't have much in the way of possessions. Relatives and friends would donate or lend things until they could get new furniture and fittings, just as people had done for them when they got married.

Elizabeth Graham remembers how excited she and her brothers and sisters were the day her parents announced a plumber was coming to fit a bathroom:

Mum had got fed up of the tin bath in the kitchen. It was so difficult with five or six young children. It was never convenient to bath them in the kitchen with people traipsing in and out. One day she dropped a bucketful of dirty water she was emptying from the bath all over a pile of clean washing. When my dad came home she shouted at him that she'd had enough, we had to have a proper bathroom.

I think she was influenced by our grandma who had a bathroom at the new house she'd moved to at Allerton. We used to go and visit and use it, it was wonderful. Anyway, when we found out we were getting this bathroom we were overjoyed. 'Don't go showing off to the other kids,' mum shouted as we ran out into the street, bursting with the news. She knew what we were like. And sure enough, every child in the street knew about it by nightfall.

As we read earlier, some children would be sewn into their clothes to keep them well insulated during the winter. Early morning was usually the coldest time in the house, before the fire had been lit. Julie Dillon's mum would always go downstairs before the rest of the family got up and light the fire. If the children woke at the same time, she would tell them to stay in bed until the rooms had warmed through.

In the evening, a shelf from the warm oven could be put into beds to warm the sheets and blankets. Another trick was to tip the remains of the living-room fire into a bucket and carry it upstairs to the bedroom, where it could give out the last of its heat.

Another strong memory is that most housework was carried out by women. Many had a weekly routine they had to stick to, simply to ensure the smooth running of the house. Washing was done once a week, usually on a Monday, and it was a task that could take all day.

Some houses had a settpot boiler for cleaning clothes. It would normally be in a cellar out of the way and took up a lot of room and filled the place with steam. People knew it was washday when they walked into a house by the steamy smell and atmosphere. Molly Carter recalls that they always had what she calls a 'make-do-and-mend' lunch on washdays, something like bread and cheese, because their mother didn't have time to do a proper meal. Her mother always seemed tired on washdays and would say to their father that if he thought working in the mill was hard he should try washing for a whole family in a small house.

Even the youngest children in a family would have chores to attend to before they could go out to play. Common chores were running errands to the corner shop, dusting and polishing, taking rubbish to the outside bins and sweeping the yard.

## The water was never wasted

In any family that had more than one person working in the mill, an early start to the day was more or less guaranteed. Some families would prepare breakfast and set the table the night before so that everything would be ready for when they came down. This allowed them to sleep in for as long as possible. While Harry Fuller was at school, he remembers hearing his mother leave for her early shift at around six. She used to come back at eight to get the children ready for school and give them their breakfast and then they all set off together.

Doreen Brook was one of six children and, when she was young, her father worked at Shaw's Woolcombers. She hardly ever saw him because he always worked nights. When they were coming home from school, he would be getting his packed lunch, or 'snap' as they called it, ready to put in the bag on his bike. She saw most of her father on a Sunday and, because she was the only girl in the family, she was the one that was taken with her dad to see her Aunt Maggie.

Like many people involved in the textile industry Doreen's

father had talent, but because of circumstances he was never able to develop it. He could play the piano by ear but, because the family were forced to sell their piano for food when times were hard, she never actually heard him play.

His father had an allotment across the road from the dye-works and he grew wonderful crops with few diseases, considering the chemicals in the air. But unless the cauliflower leaves were tied closed, the vegetables would be grey because of the pollution.

If people lived close enough to the mill they would often go home for lunch, rather than staying at work. Joan Holmes has clear memories of a day when she got a bit too comfy at home during her lunch break:

> I ate my sandwiches and had a cup of tea as usual, but I came over all sleepy and snuggled down in the fireside chair, telling myself I'd just close my eyes for a minute.
>
> I woke up a bit later when I heard some kids playing out in the street. I looked at the clock in horror and saw I was supposed to have been back at work half an hour earlier. I've never run down the street so fast. When I got back in, I told the overlooker Harry that I'd felt ill and wasn't sure whether I'd make it back. He looked at me suspiciously but never said anything about it as I wasn't usually late. I never let myself get too comfy after that, it was just one of the kitchen chairs for me.

Many families who lived in streets of older housing, like those around Manningham Mills, had to share a toilet with two or three other households. There was a cleaning rota and families took turns to provide squares of newspaper, the standard substitute for toilet paper.

On bath night, for those without a bathroom, large amounts

of water had to be heated to fill the tin bath. For this reason, remembers Kim Andrews, the water was never wasted:

> My sister and me used to go in first when we were quite tiny. Then when we were done, my mum would lift us out and my older brother would have a bath while we were upstairs getting ready for bed. Dad had to be last of all as he was always the grimiest from working with machinery and no one wanted to go in the water after him.

In the 1960s, it became more common for groups of older teenagers to live together if they were all working. Before this, girls especially were expected to live at home until they got married. It was almost an unspoken rule that, because the family had looked after you when you were growing up, once you were earning a wage you had to repay them.

Julie Dillon remembers the scandal that arose the first time she mentioned at the dinner table that she was thinking of moving out to live with a group of girlfriends at a boarding house in Manningham:

> Mum was in tears and asking what she'd done wrong and my dad looked really hurt, saying I'd broken up the family. My mum seemed more worried about what the neighbours would think about why I was leaving home and was worried what they'd think if I wouldn't be going to church anymore.
>
> To be honest though, the place we they stayed in was actually very strict. It was a big old terraced house and we were expected to be in by ten every night. I remember thinking I was allowed to stay out later than that at home. On top of that, no friends of either sex were allowed to visit, even in the daytime. If you asked

to stay out after ten, the landlady would tell you that you couldn't have a key and you'd have to stay somewhere else for the night.

She didn't even like people getting letters and would tut as she put them next to your plate at breakfast. You were expected to strip your own bed and vacuum and clean your room. There was a weekly inspection by the landlady when she even looked inside the wardrobe and cupboards, saying that she was searching for food as we weren't allowed anything in the rooms.

There was also a ban on smoking except in the house lounge, which everyone shared. There was a television, but it was always one of the older men who chose the programme that was on. If we wanted to smoke outside the lounge, we used to lean out of the dormer window in one of the attic rooms. It was quite nice doing that actually, looking over the rooftops at the rest of Bradford.

# 8

# street life

For many people, home life and street life were inextricably linked. Nowadays, a lot of people don't get to know their neighbours well and live far away from relations. In close-knit communities, such as those around mills like Drummond's and Illingworth's, neighbours were people you could turn to in times of trouble. Although there may have been few secrets and little privacy on a street of closely-built houses, anyone who was in difficulties could be assured of a helping hand.

One of the things that forged community spirit was the experience of pulling together in times of trouble. In areas where a mill employed a big percentage of people, the whole community was affected if there was a downturn in business. Even local shops would experience a drop in trade as people had less to spend.

While big events such as weddings and street parties involved everyone in the neighbourhood, there was plenty of daily contact as well. Summer evenings were a perfect time to see everyone in the street. Howard Rudge spent much of his childhood on White Abbey Road near Drummond's mill where his father worked as a labourer:

> In the evenings, if it was fine, everyone would be outside. The mums would have finished their chores and most of the dads would be home from work. They'd sit out at the top of the steps, watching all of us kids play. Most of the women would be cradling a cup of tea in their hands and some would bring out kitchen buffets to sit on. Sometimes a game of rounders would start and the

ball would invariably get knocked near to one of the parents and they'd throw it back. Then a game would start and there'd be everyone, young and old, joining in.

If a window got broken on the street during a game of football, it was up to the family of the child who'd broken it to pay for the repairs. There were some right to-dos over it, I can tell you. All of the kids would deny they'd been the one to kick the ball as they didn't want to get the blame and get into trouble at home. There was a man on the next street who always had a pane of glass ready to put in for a small price. All the windows were the same size, so it wasn't a problem, but even so, no one wanted to pay out money if they didn't have to.

One thing people didn't mind paying out for was a street party, an event that could become particularly competitive, remembers Hilary Simpson:

We held a party for Queen Elizabeth's coronation in June 1953. The weeks before, there were all sorts of ideas and recipes in magazines like *The People's Friend* for holding your own street party.

One of the older women started the ball rolling by going down to each house on the road and asking for a donation. I'd have never had the nerve to do that, but she collected a couple of pounds, which seemed a lot. Then it was just a case of beg, steal or borrow to get everything we needed together. Some of the girls at the mill who lived on the other street were always sneaking up behind us in the canteen, trying to listen into what we were planning. It was a matter of pride that your street would have the best party, everyone would brag about it to people who lived on another road.

On the day, we started with a party for the little ones. We'd got some lengths of cloth from the mill to run along the tables we had lined up down the middle of the street. Some of the material was a bit frayed, but who was looking? And we'd all got together to decide who was bringing what in the way of food. Well, the look on the kids' faces as they came out, you'd think they were at Buckingham Palace itself. I don't think the poor little beggars had ever seen so much food in one place. They were all a bit shy to start with but they soon dug in.

Later, once they'd gone to bed, the older children and adults had their own party, with bottles of beer kept cool in barrels and dancing from a gramophone. And for once, no one could complain about the noise, because all of the neighbours were involved.

## Good neighbours

Many of those interviewed have memories both good and bad of life with their neighbours. A lot of people spent much of their lives in the same area and so saw neighbours' children grow up to have their own families. Molly Carter recalls the respect children had for their elders:

Of course, people moved around, but it seemed like there was a core group of neighbours that were there all the time you were growing up. You saw them at church, when you were queuing in the shops; you listened to your parents talking about them at meal times. But as children we never called any adult by their first name, it was always Mr or Mrs whatever. Or if your mum and dad were particularly friendly with them, you might call them auntie or uncle. It would have been cheeky to use first names.

John Pashley lived in East Bowling where many of his neighbours worked nights in the Edward Ripley's dyeworks, which was believed to be the biggest dyehouse in the world. He remembers that children had to play very quietly in the street, so as not to wake neighbours who'd come in off the night shift. If the games were getting a bit noisy and riotous, the children would be sent to play in Bowling park and let off some steam without bothering anyone.

The antics of children could cause problems between neighbours, but such misunderstandings were usually limited to children from other streets. Hilary Simpson was being bullied by a gang of girls on a street that she had to pass on her way to Princeville school. Every day, they would lay in wait for her and grab her hair or try to snatch her packed lunch out of her hand:

> It never occurred to me to go another way. But I was getting more and more scared of passing these girls. And of course they knew they had power over me because they could see I was afraid.
>
> My mum found me crying in my bedroom one morning because I could see them waiting at the end of the road. She walked me to school and I was scared she was going to say something to these girls and make things worse, but she just held her head high and walked past. I went into class as usual and when I got home that night, she told me she'd been to see the headmistress about it. I was mortified, you just didn't do that, I thought. I'd be the laughing stock of the school. I still got some stick every time I went past the girls, but they were a bit calmer after that.

Diana Roberts found that women who were seen as a little bit different to others on the street could become victims of gossip.

Often single mothers or women who were seen as being too friendly to men could be ostracised from the rest of the group. Diana remembers when she was younger there was a girl called Clare on the street, who became the unfortunate victim of gossip:

> This girl must have been about nineteen or twenty and I thought she was beautiful. She had painted toenails and an ankle bracelet, which I tried to copy using a daisy chain. I could never understand why my mum and her friends talked about her in a bad way. One day, I said to mum that I wanted to be like Clare when I grew up. 'What, and have a baby before you're married?' she shrieked. I hadn't even realised she was pregnant. She moved out before the baby arrived but they had a field day talking about her for weeks.

When Molly Carter went on to have her own family, she moved from Girlington to Lidget Green and soon formed a close friendship with three of the women on her street:

> Before long, I was part of their 'gang' and we used to meet up every day. We used to do our housework in the morning and then in the afternoon we'd go into each other's houses for a chat and take it in turns to provide tea and biscuits. We'd sit and talk until it was time to pick up the kids from school. We even took it in turns to drop off and collect the kids, so that the other two could have some free time each morning and afternoon. My husband used to wonder what we found to talk about all day but all we needed was on that street, we were never short of a bit of gossip or news.

Dee Rogers was so close to her next-door neighbour that they often even shared the same evening meal:

I had this pal, Veronica, and we used to do our shopping together before we started work at two. We went shopping nearly every day because there were no fridges or anything like that for storage. Anyway, we soon worked out that if we were buying in larger quantities it was a lot cheaper. So we'd ask the butcher for enough for both of us and then both families would be eating the same meal that night.

When it was the end of the week and we were particularly skint, we'd make what we used to call blind stew. It was called that because it had hardly any meat in it, but you'd pad it out with vegetables and barley to make it quite thick and I suppose it was nourishing as well. It was the sort of thing you could leave cooking on a low light for hours and it made the house smell gorgeous when you walked in.

In terraced housing, the walls could be very thin and there was, in consequence, little privacy. Evelyn Pearson remembers hearing neighbours arguing and sometimes shouting after each other down the street. But she quickly learnt that, in the majority of cases, people made up quickly and forgot their quarrels. Sometimes she was upset if she heard her parents arguing, usually about lack of money, if her dad had been to the pub. But then the next morning at breakfast they'd be fine with each other.

Evelyn remembers visiting an elderly aunt in Lucy Street, which was off Ingleby Road and has now been demolished. It was the 1950s and she distinctly recalls seeing a solid brick wall running down the middle of the street. Her aunt told her that some of the neighbours were at war with each other and wanted a partition to define who could go where. She has no idea how long the wall stayed there, but it is one of her clearest childhood memories.

People may have been quick to flare up into arguments at times, but neighbours would also help each other if someone was ill. Some women would be well known for their knowledge of medicine and were trusted more than a doctor. Joan Holmes remembers her mum staying up all night sometimes, looking after someone who was ill:

> She didn't get paid but, after it was all over, the people she'd helped would bring round some jam or something.
>
> One night I remember we were all sitting round the fire just before bedtime. There was a knock at the door, really loud banging, and mum and dad looked at each other in alarm. It turned out to be a young boy from down the street. He said his dad had sent him because his mum was having a baby. Mum didn't say a word, just grabbed her coat and ran after him still wearing her slippers. She was away for a long time but, next morning, dad told us there was a new baby on the street and we could go and see it. I stepped into this bedroom nervously. I'd never seen a new baby before. This lady was sitting up in bed, the baby was lying next to her in a drawer that had been taken out of its chest. It was all done up lovely with baby blankets, just as nice as a cot, it was.

Before the NHS was established in 1948, there would be a bill every time a doctor was called, which was one reason that people looked to friends and neighbours first. If a doctor was called, the bill could usually be paid in instalments and a collector would come round for the money each week. Some doctors would even waive their fee in circumstances of extreme poverty, but this was at their own discretion.

Just like a birth or marriage, a death would be treated as a community event, with everyone keen to pay their respects to the deceased. Particularly during the early part of our period, a body was left in the house until the day of the funeral rather than being taken to an undertaker. People would come in and pay their respects, removing headwear as they entered the house. A common superstition was that all mirrors in the house of the deceased had to be kept covered until after the funeral.

The bereaved family was expected to provide a funeral tea after the service. This could either be arranged by friends and neighbours, or by the undertakers. Often the whole street would draw their curtains at the time the funeral procession was passing. Men would doff their caps as a hearse passed, even if they did not know who had died.

A big fear was not having enough money to pay for a decent funeral. For many, the ultimate shame was a pauper's funeral, without any of the trimmings such as a hearse, wooden coffin and funeral tea. For this reason an insurance policy, based on a modest contribution that was collected weekly, paid out in the event of death.

## Corner shops

Having a good relationship with local tradespeople could pay off time and time again when shopping for the family. The relationship was a two-way one, as the shopkeepers needed custom and shoppers liked to get a bargain. Favoured customers would often be treated to free or cheap goods that had been damaged, such as biscuits or chocolate.

Harry Fuller's father had a flourishing allotment alongside Ingleby Road and was able to use the fruits of his hard work to barter with local shops:

He was clever with the system he had. He always ended up with more vegetables than we could eat at home and so he took them around the local shops. Although the shopkeepers would be reluctant to pay cash for the goods, they would often swap them for things we needed, like meat or cleaning stuff.

Many Bradford shops accepted club checks, and Manchester Road was particularly noted for having a good supply of shops that would take the checks as payment. The check system was an early form of shopping club, with people getting a check from the club man and then repaying it by weekly instalments. An amount would also be added for interest.

Harry Fuller remembers that once, when the family was particularly hard up, he was told to watch out for the club man coming down the street. When he was getting near to the house, his mum went inside and shut the door, ignoring the clubman's knock. But, the next week, she had to find double the money and ended up owing money to the rent man instead. This taught Harry that if you couldn't pay for something, you were unlikely to make it up the next week.

Elizabeth Graham was sent round to the baker and the grocer during the 1930s when her father was out of work to ask if they could spare anything for a needy family:

Because mum had been a good customer in better times, I'd be given vegetables that had gone off or buns that had been dropped on the floor for nothing. I suppose people felt sorry for a child. I'd only have been seven or eight.

I remember once I was given a bag of iced buns that were going stale. On the way home I counted them and saw that there wouldn't be enough for everyone in the family to have one each. Without thinking, I greedily ate one of the buns before I got home, as I was so worried I wouldn't get one. But then, after tea, mum said that she didn't want one and everyone else could have her share. I felt so guilty, but then again, not that guilty that it stopped me eating my second bun!

Shopping for clothes was done on a much less frequent basis than shopping for food, if it was done at all. Nevertheless, many parades of shops and some of the city-centre streets had tailor's shops. A popular tale is that Bradford tailors would dread a husband and wife coming in to buy a suit because the chances were that one of them worked in textiles. This meant that they would spot even a minute flaw in a piece of cloth and couldn't be fobbed off with inferior materials.

Miss Watts has lived in the Manningham area all her life, and spent most of her working years at Lister's. She was recently invited to open an exhibition at the newly reopened Lister's mill, just after it was converted into flats. She has very clear memories about the type and varieties of shops that once existed around the mills to serve the local community. She entered the mill as an errand girl straight after leaving school. Her responsibilities included collecting the lunch orders and a number of similar tasks, such as collecting the boss's dry cleaning. She then worked in the dress-silk department weaving the fine material that was also used as a luxurious lining for jewellery boxes.

She remembers that Oak Lane had a particularly fine parade of shops of every description. These included a grocer, a confectioner and a Chinese laundry, where she used to take her boss's collars to be laundered. There was a popular pork butcher where people bought crackling to eat for lunch. As part of the lunch orders, she would be asked to collect buttered teacakes from Fieldhouse's the baker and hot pork pies from Hanson's, a grocer. There were two kittens in one of the shops named 'damn and blast', because people were always tripping over them. On nearby Lilycroft Road there was a fish-and-chip shop, a post office and a wine shop.

When Kim Andrews got a Saturday job as a teenager at the Cooperative store on Legrams Lane, Lidget Green, she quickly became the most popular person on her street:

> I couldn't believe how many people came in to see me. There were a few of the lads on the street kept knocking on the shop window and teasing me. Then my mum kept coming in to see how I was getting on. I was so embarrassed. There I was, trying to act like a grown-up working girl and half the neighbourhood was out sharing my big moment. But that was what it was like round by us, everybody had to get involved. I don't mean that in a nasty way, but you have to think, there wasn't a lot going on usually and so even something like that was a big thing.
>
> The job itself was quite good fun, but you were on your feet all day. I'd imagined myself sitting behind one of the tills, with my nails all painted up and my hair styled, but no such luck. I started off stocking the shelves and cleaning up if someone spilt something on the floor. Very glamorous! You'd get people trying to steal as well, we were always told to keep an eye out for that. I wore a long blue house-coat-type uniform, which I hated, but actually, it saved your clothes getting wrecked, so I suppose it was a good thing. It's just at that age, you're all out to impress; you never knew who might come in to see you.

The pawnshop was a necessary evil and there were many pawnbrokers in and around Bradford. People got paid weekly and, once they had accounted for the basics like rent, fuel and food, would sometimes have to put items such as the husband's best suit into the pawnshop to get a little extra money and then redeem them at the end of the week.

Some of the men didn't even realize that goods were going to the pawnshop. Women would pawn valuables after the men had gone to work or send the children with them. In some cases, they would even pawn their wedding ring. Once people got used to pawning items, it could be a vicious circle. It became harder to manage without the pawnbroker and it was sometimes impossible to raise the money to buy precious possessions back.

## Characters

As well as the traditional shops, a range of tradespeople sold goods on the streets. The first person that people would hear at the beginning of the day would be the knocker upper, who was paid to wake people who started work early. As we will see in the next chapter, the ice-cream man was a favourite with children. Some people also remember a pop man who came around the streets selling ginger beer in stone bottles.

The rag-and-bone man's call was a familiar sound, remembers Joan Holmes:

> You could hear him shouting 'rag and bone' from streets away. We were always a bit frightened of him, people used to say he lived on his own in a cave up Thornton. He always gave children a balloon if they took out some rags for him to have. I remember once my brother was desperate for a balloon, but he didn't have anything to give. So the daft lad took his bike and handed it to the rag-and-bone man, who must have thought it was his birthday.

A few minutes later my brother started crying, realising he wanted his bike back and ran in to tell mum. The three of us had to chase the rag man for several streets before we caught up and I remember mum had a real job convincing him to give the bike back. As for my brother, he was banned from riding that bike for a week.

The street characters who stuck in people's memories were those whom children were afraid of. Evelyn Pearson firmly believed that a witch lived in one of the back-to-back houses behind her street. Because the house was usually in darkness, with an overgrown patch of garden and all the windows covered, children would say that a witch lived there. They'd dare each other to walk up to the door or throw a pebble at the window and then run away screaming.

Brian Fraser had a real fear of an elderly man on his street, who he later found out was a war hero:

There was a chap lived a few doors down who had a wooden leg. He'd walk down the street with a really bad limp and if you got under his feet, he'd growl at you really fiercely. He always wore a grey bowler hat that hid most of his face.

I told my dad I thought he was a nasty-tempered old man and he went mad. He said that Ernest had given up a lot for his country; he had fought in the first world war and I should never say anything bad about him. If it wasn't for him, said dad, the country would be ruled by Germans. It didn't mean much to me at the time, but when you think back, you realise how you could misunderstand people like that. He was probably in pain and didn't want a load of kids tearing past him.

On icy mornings, men would be employed to put ashes on the pavements so that people didn't slip. Casual seasonal work like this was often given to men who had signed on as unemployed. Brian Fraser's father used to pray for snowy weather when he was laid off as a spinner because it meant he could get a few hours work shovelling snow off the roads. Other times he worked tarring the cobbled roads. People used to follow the tar wagon to breathe in the fumes, as it was believed to cure coughs and colds.

Perhaps the least popular, but most necessary, street job was that of night-soil man. These men came around every few months to empty the middins, which contained sewage and ash pits, and were used for rubbish that couldn't be burnt on the fire. This unsavoury job was done at night, to minimise the unpleasantness for householders. A more pleasant job was carried out by the lamplighter, who appeared each evening before electric street lighting was introduced, to light the lamps.

## Pea soupers

Pollution was a fact of life around a mill, but it wasn't something that bothered most people. Often a trail of smoke coming out of Lister's chimney could be seen for miles around. Because Bradford lies in a valley, the smoke and pollution would rise out of the bowl-shaped land, creating a spectacular scene from higher ground.

David Briggs remembers the 'pea soupers' – which is what smog was called – and says that on particularly bad days, soot and grime would stick to clothes and faces. He attended Undercliffe school on Undercliffe Street, said to be the longest and steepest street in Bradford. He remembers enjoying the view at the top of the street: a spectacular panorama of the whole city.

On a foggy day, the view down to the city was completely

obscured and only the hundreds of mill chimneys were visible. As the sun came up it burnt off the mist and gradually the city could be seen again. Even on a clear morning, there would be lots of smoke as people lit their fires before work.

The thick pea-soupers caused problems both by day and by night. The fog was sometimes so thick that a hand couldn't be seen in front of the face. Andrea Ratcliffe remembers conductors having to get off their buses and guiding them on foot because the driver couldn't see where he was going. On a similar theme, Emily Hoyle once had to get out of the passenger seat of her car at Sticker Lane after an evening out and direct her husband right back to Lidget Green using a torch because the fog was so bad. She didn't feel afraid because she knew there wouldn't be that many cars on the road late at night.

# 9

# chilঠhooঠ

Some of the most vivid memories were of childhood in Bradford. Growing up varied from household to household but, for many children, the mill was part of their life for as long as they could remember. Whether it was running an errand for someone in the mill, playing in the mill yard after hours or hearing the buzzer every working morning, children living in the proximity of a mill were well aware of its existence.

## The next step in life

At one time many families were proud that their children followed them into the mill. After all, it could provide a secure environment working alongside people they had known all their life, and there was always the chance to earn overtime or learn new skills by moving to a different department. For many youngsters taking a mill job was as logical as going to school at the age of five. It was just the next step in life.

Dee Rogers remembers a careers lesson a few weeks before she was due to leave school, in the 1930s. The teacher had told them that an adviser would be visiting to talk about job opportunities and how to shine at an interview. 'We were so excited. My pals and me talked about the sort of jobs we were going to do. Perhaps we were going to be working in Woolworth, a baker's shop, even in a café in the market. We chatted for ages and got so excited.'

But the visit was to be a big disappointment. 'A stern man in a grey suit stood in front of the class,' remembers Dee, 'and introduced himself as one of the directors of the local mill.

And for the next half-hour he droned on about the different jobs available in the mill and how to choose which one you might like. And that was it, the end of our career talk.' When Dee tried to speak to her teacher at the end of class, and mentioned one of the other places that she wanted to work, she was told that working in the mill was secure and that she should be grateful to get work near home.

After the second world war, there was a new attitude among parents about children following them into the mill. Many had experienced higher-paid war work elsewhere and consequently wanted something better for their children. If a son or daughter managed to get a job in a shop or an office, it would be the talk of the street. The decline of the textile industry was also a factor: people found themselves forced to do part-time work instead of the full-time employment they'd been used to, and they realised that mill work was no longer a secure occcupation.

Doreen Brook's memories of a childhood in a mill community inspired her to write a poem entitled 'Child Labour', which she shares here:

*The Dark Satanic Mills are no more.*
*The mill girls do not clatter in clogs*
*Down the alley ways and ginnels of long ago.*
*Seven or eight mouths to feed on meagre wages.*
*The dripping bread for the impoverished hungry souls.*
*The little bairns who peed the bed, head to foot, six to a bed!*
*Lass wed lad and not a penny to spare.*
*No pill to take to cut the number down.*
*Half-time work, Mass and then school, and then mill.*
*A pint pot of tea was like manna to their souls.*
*Sometimes unshod feet ran over cobbled stones.*
*When ya mam tucked you up, all nine or ten of you in your bed*
*Sugar and cocoa in newspaper were the only sweets you knew.*

*The knocker uppers have long gone.*
*'On the never never' is still around in the credit card of today!*
*All the mills have now all gone.*
*They have become the lust apartments of the yuppie set.*
*There's roving and twisting of a different kind.*
*The mills have lost the sound that I remember.*
*No one is flitting as before.*
*From six to six they all worked with a deafening roar.*
*You took your tea mashing and wore pinny and shawl.*
*Overlookers shouted and bullied the girls.*
*'Get that end up!' and 'Doff that load!'*
*Now I go up the steps of the mill made all new.*
*It's become an exhibition place, and all that jazz!*
*But I swear I can hear the clogs and clatter of long ago,*
*And all those shawl-clad heads are back once more,*
*Hurrying so they don't get quartered if they are late.*
*I smell the oil and the grease of the looms now gone.*
*Dear blessed Time, Shalom.*

For those children not old enough to work in a mill, even just passing by could mean being drawn into the community. Mary Hanney remembered passing when some of the workers were leaning against the mill wall having their break. As she passed, one of them began to whistle 'Pop Goes the Weasel'. However fast or slowly she walked, they would time the tune to the rhythm of her walk.

Other workers resting outside would be drawn into a game of football or even marbles, if children were playing nearby. 'We always spoke to folk in the streets around home,' remembers Molly Carter. 'You were told not to talk to strangers, of course, but the people in and around the mill, we saw them every day and they seemed like part of an extended family.'

Those who went into the mill on a message often had a

strong memory of the environment, even years later. Evelyn Pearson used to take sandwiches to her father during her school lunch-break:

> My mum would make a pack-up for my dad in a little basket and I'd have to take it in. You know, I used to dread it. I must have been about nine or ten and it seemed a never-ending walk down that weaving shed towards my dad's machine. People would shout out to me and I could never hear what they were saying over the deafening noise. There was no point even trying to talk to my dad. He'd be really busy and I couldn't hear him anyway. He just used to pat me on the head and I'd run off, glad to be back outside.

Children often ran errands for people who had worked all day and didn't feel up to going out again after work. Brian Fraser remembers he once had a good little round fetching fish and chips or groceries. He'd be paid a couple of pence each night:

> Fridays were the best because nearly everyone had fish and chips. I used to love the feeling of bringing back a meal to a hungry family and the smell of salt and vinegar as I queued in the chip shop. When I'd delivered all the parcels, I'd run home with my earnings and look forward to my own tea.

Some children went out to meet their parents as they were coming home from the mill. Molly Carter remembers a whole gang of them waiting at the end of the street to hear the buzzer going off:

> As soon as people started coming out, we'd run

towards them. We were especially keen when we knew our parents had been paid and would nag for a penny for sweets. As soon as we got it, we ran straight to the corner shop, usually with instructions to get something in for tea as well.

## Games

Games usually involved things that children found or made, rather than shop-bought toys. Even in the 1960s, when there was a larger range of toys entering the country from overseas, children tended to invent their own games and fun. Pram wheels found new life as go-karts, old pushchairs or orange boxes could be racing cars, hilarious if you pushed a friend from the top of a hill.

Selling things was another pastime for enterprising children hoping to make a few bob. Molly Carter remembers making perfume from wild flowers she and some friends found growing on waste land. They put the petals into a tub of water and stirred, sieving it out to make a smooth liquid. Enterprising they may have been, but when they took the potion around to different houses they were astonished that no one wanted to buy it. After several tries, they gave it to Molly's baby sister and giggled as she drank the whole thing.

On a street of back-to-backs, Molly recollects, a patch of flowers would be a rare sight:

All we saw were dandelions. And we'd never pick them as the older kids told us they'd make us wet the bed. I remember once finding a garden full of daffodils while I was wandering around quite a way from our street. Well, I'd never seen anything so bright and pretty. Without thinking, I picked some for my mam and took them home to her. 'Where did you get those?' she said as I presented them to her. I remember feeling so put out. I'd gone to the trouble to pick a bouquet for her and all she wanted to know was where I'd got them.

Young lads were less concerned with flowers and perfumes. Those lucky enough to live near a mill yard had a ready-made playground and a never-ending source of games. Often, they'd find abandoned tins of grease or paint and poke around in them with sticks, 'painting' the walls or floors around. Bales of stinking old wool or discarded fabrics were used for games and made into materials to create dens.

Ernie Blake remembers that most parents told their children not to play in the mill yards. 'It was a sure-fire way to make us more tempted to go,' he chuckled. 'We'd wait on the corner until the workers had left and then, when the coast was clear, sneak in and start our games. On Sundays, when the mill was shut all day, we'd say we were going for a walk and spend hours messing around there. The dirtier we got, the more we liked it.'

Ernie remembers they'd pretend to do deliveries using whatever 'vehicles' they had. Other times, they pretended to be workers coming in and out of the mill. They copied what the adults did, leaning against the skips pretending to have a cup of tea. On the rare occasion that a skip was left outside, it was considered a real bonus and was borrowed for rides or to use for a variety of games. As long as they returned it, said Ernie, they felt no harm was done.

The perimeter walls and gates of the mill would also be used frequently for games at weekends when the mill had closed. The older boys would climb on the wall and run the full length. Sometimes they even jumped across the gap provided by the gates. Smaller children would use the walls as shops or houses or play houses in some of the outbuildings. Because some of the walls had cut glass on top to deter burglars, Ernie's mum was always worrying about her children falling and constantly warned them about the dangers.

## Tin-can squat

At one time, cigarette companies issued series of cigarette cards, popular with children and adults alike. These were an early form of 'pester power', as children would ask parents to buy certain brands of cigarette, so they could get their hands on a set they were collecting.

Children swapped with each other to get the full set, usually around twenty cards. There were different illustrations on the front of the cards including people of all nations, ships, sports stars, trains and animals; the other side of the card was full of facts about the subject. Their popularity was such that groups of children hung around corner shops and asked people for the cards as they came out with their cigarettes.

Most people remembered playing in the street, rather than in each other's houses, whatever the weather. Popular games included whip-and-top and marbles. In another game – skipping with a rope tied to a lamp-post – they took turns to hold one end of the rope and tied the other round the post. If someone was lucky enough to have a bike, the owner was able to lend it to other children in exchange for a few sweets.

Tin-can squat was a game in which one person kicked a tin can as far as they could and someone ran to bring it back. While the can was being retrieved, the others ran away and the purpose of the game was to find them. Hide and seek was also popular and inevitably ended up spilling into areas where children had been told not to go.

Sometimes they'd hide in gardens or even in churchyards during hide and seek. Molly Carter remembers hiding in the entrance to a railway tunnel even though this was strictly for-bidden by her parents. 'I knew I shouldn't have gone in, and all the time I was wishing someone would come and find me before I got caught by a grown-up,' she said.

Ernie Blake went fishing in Lister Park, using some old equipment he and his friends had found. They would dig up worms to use as bait. It would be difficult to fish because of all the people who were going past, the swimming in the lake and the boats. 'We hardly ever caught anything, but we felt so grown-up,' said Ernie.

Most neighbours were tolerant of children playing on the streets. Some, though, took a dim view and were disliked by the gangs of children and sometimes even picked on. It wasn't unknown for people to throw a jug of water at children or, if a ball went into their yard, to puncture it or refuse to give it back. Molly Carter remembers her dad always telling her never to involve parents in an argument between children. His reason-ing was that the children would make up after an argument, but the parents might still not be speaking years later.

Sometimes children would put on shows for their parents. They would take it in turns to perform turns they had seen on variety shows or pantomimes. 'We were so cheeky,' said Evelyn Pearson. 'We'd go round knocking on doors and ask people if they wanted to see our show, knowing very well that they'd probably give us a drink or some sweets at the end of it.'

After seeing *Mary Poppins* and she and her friends decided to draw chalk pictures on the pavement just like in the film. 'We drew landscapes and pretended we were jumping into them like Mary Poppins did. I was upset when it rained that night and all we were left with were a few smudges of chalk.'

Another chalk game was called 'follow the arrow', and it was mostly played by older children. One group would chalk arrows onto walls and the others had to follow where they led. Often it would go on for miles and sometimes two arrows were pointed in opposite directions to throw the group off the scent.

Younger children were allocated a defined territory by their parents. It usually extended to the end of the street. 'We usually played where we were supposed to,' said Brian Fraser. 'But sometimes, someone would dare you to go further. But you knew very well where you could play and most of the time stuck to it.'

He remembers running away from home after an argument with his mum. He marched off down the street but then remembered he was only allowed as far as a red-painted house and so he stopped there. After a few minutes, he wondered why no one had noticed he was missing. 'I went back to the house, and there was my mum at the kitchen table, making some parkin. She asked if I wanted to lick out the mixing bowl and that was that. She hadn't even noticed I'd gone.'

Despite all their adventures, accidents were few and far between. Because they were so rare, they stuck in people's minds when they did occur. Molly Carter remembers a boy being drowned swimming in a mill dam in the 1930s and parents used to warn their children not to go near after that. It did frighten them, said Molly, as did an accident where a boy was knocked down by a lorry outside his school. Fortunately, he wasn't killed but was in hospital for a long time.

On the side roads, traffic wasn't a problem. Up to the 1950s, cars were a rarity and delivery vans usually visited shops early in the morning. It wasn't uncommon for children to set up games in the middle of the road and not be disturbed. By the 1960s, more families had cars and it was no longer as safe for children on the streets. The main roads had always been busy, but now cars started to come onto the side streets as well.

Evelyn Pearson remembers the ice-cream van came to her street once a week, on Sundays, and that the whole family would listen out for it:

We'd have our money ready and as soon as the van turned into the street, you'd see kids running out of the different houses. Some would be in their pyjamas, as it could be as late as eight at night, but no one wanted to miss out.

The ice-cream man would always take your money before handing the ice cream as some children would take the ices and run off without paying. The best was a 'special', which was an ice cream with two ice-lollies stuck in it and sugar treats all over it with sauce dribbled on.

Corner shops normally had a display in the window, and some would put in special items at Christmas and Whitsuntide. Dee Rogers also remembers the pawnshop being a place where children would stare wistfully at the goods on display in the window. The pawnshop once had a pair of black patent shoes that she was particularly fond of:

It never occurred to me that they were second-hand, I just loved them. They had a strap around the ankle and a little flower shape cut out of the front. I pictured myself wearing them for school and everyone crowding round me to look at them. When I told my mum she was horrified. 'We might not have much money, but you're not having second-hand shoes', she said. It was a matter of pride to her that we'd always had new shoes, even if they were sturdy ones that I didn't like.

## Making mischief

Mischief was a big feature of street life and included things like putting string on door handles, or knocking on doors and running away. Dares would include jumping-off walls or running through gardens, even knocking on windows for those brave enough. This was tolerated by neighbours. 'You knew what you could get away with,' said Brian Fraser. 'There were houses where you'd get a clout round the ear for annoying people and you tended to avoid those. But as long as you weren't damaging property, people were good humoured about most games. No doubt they remembered what they used to get up to when they were young themselves.'

Another popular prank was played at those shops where sweets were put in glass jars and weighed out. A child would ask for sweets from the top shelf and the shopkeeper would climb up and weigh them out. Then the next child would ask for something else from the top shelf, pay for it and then the next child would ask, making the shopkeeper go up and down the ladder several times.

Bonfire night was a highlight and children looked forward to it for months. There would often be dozens of fires on the streets and, when one burnt out, people moved on to the next one. 'Chumping' (getting hold of wood to build the fire) could begin as early as September. Competition for wood was fierce and it was often pinched if left on the street. Brian Nelson says that building a bonfire was not something that was taken lightly:

We took our bonfires very seriously. The girls wouldn't bother much, but among the boys, having the biggest pile of wood and guarding it seemed like life or death at the time. There were always gangs from the other streets waiting to take wood from you. I used to persuade my dad to let us store the wood in the cellar until just before bonfire night. He'd moan that it was a fire hazard, but he always gave in and let us keep the wood there.

## School days

School days were usually spent with children from the local area. Even if children attended a church, rather than a state, school, they would still know many of their classmates from home. The only exception to this was if someone won a place at a grammar school by getting a scholarship, an achievement that was not without its own problems as Kenneth Nelson explains:

> There was a lad on our road went to the grammar school. I always thought it was the worst thing his parents could have let him do. He'd walk down the road with this green cap on and kids running after him and shouting. No one wanted to play with him after school, he was different to us. And the worst of it was, he actually ended up working on the roads for the council, a job he could have got if he'd gone to school with us.

Even if a family did want their child to go to grammar school, money was often a barrier, even if the place was free. Bright children often weren't allowed to sit for their scholarship because their wages were needed at home. The only way to continue in education was to take evening classes after work.

Sometimes children were kept off school to look after parents or siblings who were unwell, or to make extra money when times were hard. 'If mam had got some outwork and we were particularly skint, we'd all have to pitch in,' said Dee Rogers. 'So mum would write a note for school saying I was off sick with a cold and instead I'd be at home getting stuck into whatever work we were doing at the time.'

Although most children in a class would be from the same area, and so in a similar income bracket, there were always a few children who were really deprived, particularly during the

1930s. Some schools offered free breakfasts to children and even cast-off clothes.

Many classes contained over fifty children of various ages and abilities. Molly Carter remembers that teachers could be quite fierce with pupils. She remembers a boy in her class being slapped in the face hard for not paying attention during class. 'Other teachers could be lovely, though,' she said. 'If you had worries, or if you were behind with your homework, they'd tell you not to worry and help you out. I think most of them understood that we all had troubles at home from time to time and tried their best to support us.'

At Roman Catholic schools, the teacher would pick a few children at the beginning of the week and ask them if they'd been to mass on Sunday. They even asked what colour of vestments the priest had been wearing, or what the sermon had been about, to make sure they were telling the truth.

Molly remembers the Lord Mayor and Lady Mayoress coming round the school one year and distributing toys to the children. She doesn't recall what the occasion was but has a vivid memory of the beautiful doll dressed in a sailor suit that she was given. When she was going home it was taken by a gang of older girls she knew from school and she ran home crying. Her mum asked her what had happened and went out straight away. She came back with the doll but wouldn't tell Molly what had happened.

## Sunday school

Weekends may have been an escape from lessons, but many children also attended a Sunday school, or went to church or chapel with their parents. Sunday schools were aimed at children from age four until they started work.

Whether or not they enjoyed Sunday school, there were

attractive fringe benefits. Some schools would give out a token for every week of attendance and, the more tokens accumulated, the better present that was received at Christmas. There were also summer outings, such as picnics or walks.

The choir was another encouragement for children to take part in church life. There would often be a junior choir and then an adult one for people who had started work. For many, singing in the choir was the start of a lifelong interest in music.

Kathleen Wright remembers one occasion when her church choir was asked to sing for a special service. The choir was in a special section apart from the congregation. While they were singing, one of the girls accidentally wet herself on the bench. When the song came to an end the choirmaster indicated for them to take their seats but, understandably, nobody would, and they kept standing until the service had ended.

## Chores

In a family where both parents were working, children were expected to do jobs in the house. Molly Carter remembers these could include black-leading, cleaning windows or scrubbing the front steps. 'I remember I always used to be careful never to say to my mum that I was bored,' she laughed. 'If I did, she'd always find a job for me in the house.'

Older children would take a paper round or get informal work at local shops delivering bread and groceries. Sometimes parents were reluctant to let their offspring take a job, as Brian Fraser explains:

> I nagged my dad to let me have a paper round. He told me I wouldn't like getting-up early, but I was insistent. The first time it rained, I didn't want to go out there so I asked him if he'd come with me to make it quicker.

No way would he do that. He told me if I was responsible enough to do a job, I had to take the rough with the smooth. I think I lasted just two weeks on that round.

## Christmas

Christmas was eagerly anticipated. Although money was scarce, people would make sure that their children had the best time possible. There were Christmas clubs at various shops to help spread the cost of presents through the year. You weren't allowed to take the items until you had paid for them, but it was a good way to save, as money kept at home would end up being dipped into for other purposes.

Hiding presents was a problem with curious children in the house. This was where friends and neighbours would come in, by hiding gifts in their houses. Dee Rogers remembers seeing a gorgeous doll when she was hunting for presents with her sister. She just knew the doll was for her; it was in a see-through box with a feeding bottle, bowl and spoon. She loved it so much that she decided to name it Lisa:

> Christmas came and went. I got lots of nice presents but all the time I was looking for this doll. I was almost crying in frustration when I'd opened everything. When I saw my friend Tara after Christmas, she told me about a beautiful doll she'd been given. Before I even saw it, I just knew that it was 'my' doll. I never, ever wanted to play with it after that.

# 10

# the holiday spirit

As we read earlier, most Bradford mills did not offer paid holidays until at least the 1950s. But this did not stop people enjoying holiday times either by staying in Bradford or going further afield. Whether experiencing a fair close to home, taking a day trip into the Yorkshire Dales or saving up all year for a week at the seaside, many people's memories of holidays have lasted them a lifetime.

Nowadays, most workplaces arrange holidays so that there aren't too many people off at the same time. But, in the mill, the entire factory would shut down for the holiday fortnight. The only people working were those employed to carry out the annual clean, as well as maintenance and caretaking staff.

Evelyn Pearson remembers the joy that used to accompany the Friday when Drummond's mill closed for its holiday fortnight:

> One time, I was coming along the street towards the mill just when the final buzzer went and people came pouring out. They were all laughing and talking at once, a few of the women threw their overalls up into the air and everyone was linking arms and singing 'show me the way to go home'.
>
> There was such a big crowd coming towards me that I didn't even attempt to make my way through in the opposite direction. I stood on a step and just watched them all come past, it was quite a sight. And as they walked away, you could still hear the laughing and singing even when they'd gone round the corner.

## All the fun of the fair

The summer holiday periods were called 'tides'. Three of the main ones were Bowling tide, Manningham tide and Shipley tide, named after their respective areas. Travelling fairs and shows were well aware of holiday periods and arranged to visit Bradford's parks at these times. Posters would appear in shop windows for weeks beforehand advertising a circus, with brightly painted pictures of the animals and the other attractions.

Many of the big-name circuses, such as Scott's and Smart's, visited Bradford and were eagerly anticipated. Any shop that allowed a poster to be displayed in their window would be offered free tickets. To add to the excitement, there would often be a circus parade before the big event. Harry Fuller remembers his first visit to the circus without his parents:

I was about eight or nine and I persuaded mum to let me go and watch the parade with two of my friends. She spoke to their mams and they decided it would be all right as long as we stuck together. We'd been told that the parade would be coming along Manningham Lane and we'd already decided we'd stand on a wall at the bottom of Oak Lane and so have a great view of the parade actually going into Lister Park. There must have been hundreds of people waiting, it was a great atmosphere and there were people walking about selling toffee apples, sweets, things like that. At last, we heard some music and I spotted a line of elephants coming along the road. It was so exciting seeing animals like that in your own city!

Each elephant had a girl riding on it, dressed in Arabian costume and waving to the crowd. There were jugglers walking alongside and a man handing out leaflets about the shows. As the parade passed, we ran alongside and followed it into the park. The big top was already up and the first show was that evening. We weren't going, and I found myself wishing I had tickets. But even so, we'd seen the parade so we didn't feel too badly done to.

Bowling had the reputation of hosting one of the best fairs in Bradford. The caravans and trucks involved in the fair would travel in, often straight from running a fair at Manningham. Children would crowd around to watch the rides being put up and would sometimes help the showmen set up in return for the promise of a free ride. The fair opened on a Friday night – when most people got their wage packets – and was closed all day on Sunday.

As well as the May processions held by Bradford's churches, there were also parades and maypole dancing. Many schools arranged events and the children would be invited to bring along a pram, bike or scooter trimmed-up for a procession through the streets. Kim Andrews attended St William's in the 1960s and remembers how involved the whole school was in the May Day celebrations:

A week or so beforehand, we all took home a letter to our mums and dads, asking if we could bring in something for the May procession. I was dying to take my doll's pram and my mum came up trumps with decorating it. She bought some pink-and-blue crepe paper from John Street market and she did wonders, wrapping it all around the wheels, handles and hood. She made long tassels for the edge of the pram handles and even created a matching bonnet for my doll out of crepe paper. On the big day, I wheeled the pram down to school. It was a long walk from Lidget Green where we lived, around twenty-five minutes, but I enjoyed every moment, seeing what everyone else had brought along. It's just a good job it didn't rain or my marvellous pram would have been a mass of soggy paper.

We had the school assembly and left all the bikes and prams outside, and then we came out to walk along past St William's church and back to the school. All the mums were watching and clapping us. Those sort of things stay in your mind for ever, something a bit out of the ordinary. Perhaps they weren't expensive fancy events like kids go to today, but to us, they were days to remember and I'm sure the mums loved watching us.

The family of Dee Rogers also played a part in May Day events. She remembers her father, who worked as a carter, staying up almost the whole night to get his horse and carriage ready for a May Day parade. When he came in from work he had his tea and then decorated the cart with crepe paper, while her mum made flowers out of tissue paper and pipe cleaners. They took it in turns to wind paper around the wheels and edges of the cart, to make it look as pretty as possible. They also made a huge archway, and decorated it with flowers, to go over the top of the cart. Dee's father later said that the look on the children's faces the next day made it worth staying up and missing out on his sleep. They ran towards the horse and cart with looks of wonder and fought each other to clamber on.

Bradford's Christmas lights were another opportunity for a day in town, and many families combined a Christmas shopping trip with a look at the festive lights and decorations. The whole city was decorated with tinsel, Christmas trees and fairy lights. Hilary Simpson was able to decorate her whole house cheaply with the wide variety of decorations available at Kirkgate market:

I used to love going to the market at the beginning of December. They had whole stalls given over to Christmas decorations. There were little nativity scenes, tinsel, fairy lights and baubles of all colours, some with pretend snow and crystals on them.

I always took my daughter along, she loved to help choose decorations and they were only a penny or so each. Then, after we'd got them, we'd have a cup of tea and a bun at one of the little cafés that ran along the back of the market. When we got home, she'd dash in to show her dad what we'd bought and we'd spend a happy afternoon trimming up the house.

It was lovely to see Christmas trees and lights starting to appear in the houses as the weeks went on. I don't remember any house that wasn't trimmed up, even the poorest would manage somehow. If you brought out the same old decorations year after year it didn't matter, the thing was to make the effort and have everywhere looking nice.

## Day trips

One of Bradford's attractions is that it is close to places of interest that are ideal for a day trip. Some of the places where Bradford's mill workers visited were so close by that they could be reached on foot. As far as Ernie Blake was concerned, the adventure was in just setting off. It was not necessary to go to a far-flung place, just somewhere that was different from the norm:

My wife was very good at making a big treat out of a little thing and the kids loved her for it. When the children were young, we only had a few pence left out of my wage when we'd paid the rent and bills. Going anywhere on a bus or train was out of the question, so if we wanted a day out, it had to be somewhere three small children could reach on foot.

I remembered going to Bracken Hill park at the top of Lidget Green as a youngster and we decided to try there with the family. Ivy, my wife, told the children all about where we'd be going and that we were taking a picnic and they were wide-eyed with excitement. On the day, they helped pack the bags with the sandwiches she'd made and she sent me to the shop for some lemonade powder.

It must have taken us nearly an hour to walk there, you know what it's like with kids, they stop every few yards to look at things, but it was good fun. At last we got there and the little ones raced to the playground. Ivy and me set out the picnic on a rug and she showed me some little cakes she'd made as a surprise. We stayed there hours, lying on the grass, playing ball games and just watching everyone enjoying themselves. It seemed strange looking back down into Bradford and thinking I'd be back at Illingworth's the next day; it all seemed so far away.

When I tell the grandchildren about the places we went in those days, it doesn't seem anything to them; they can't understand it. But if you were cooped up all day in a hot and noisy mill, even just getting out to a park was a memory you'd keep with you all week when you were working hard again.

For the more able or energetic, community groups and churches arranged hikes to Bradford's outlying areas. Common destinations were Haworth moor, Shipley Glen and Ilkley. The Wharfe valley, with popular places like Bolton Abbey and Otley, could be reached via the popular Shipley Glen electric tramway. The tramway was particularly convenient for workers at Salt's, as it was just a few minutes walk from the mill.

The tramway, which still operates as the oldest electric tramway in England, took people from Saltaire up to Shipley Glen. From Shipley Glen, hikers could cross the moorland towards Ilkley, Otley or to the popular Dick Hudson's public house on the road to Keighley. Dick Hudson's was a traditional destination for Whitsuntide walks and was usually packed with drinkers during the summer holidays.

Molly Carter remembers a trip arranged by one of her colleagues at Northside Mills to Harewood House on the outskirts of Leeds. She later realised that the stately home isn't that far away, but at the time it seemed a long way off and it was particularly memorable as it was her first trip after leaving school:

Everyone had been told to gather in the mill yard at nine in the morning one Sunday. They'd been talking about the trip for weeks beforehand and deciding what they'd wear. Some people said they'd already been but most said they'd only seen pictures. One lady who'd visited before brought in an illustrated guide-book, which had photos of the house and its contents as well as the grounds and the bird gardens. We all crowded round looking at it while we waited for the bus.

We all brought a picnic and mum had let me prepare my own. She'd even let me get a packet of crisps from the corner shop, something we could never usually afford to do. We got on the bus when it turned up and started singing songs to make the journey go faster. After a while, some of the younger lads sitting on the back seat started getting rowdy and showing off to the girls, throwing things and swearing. The driver stopped the bus and came storming up to them. He said that if they didn't stop messing about he'd turn round at the next roundabout and we'd all be taken back. Everyone glared at them and they toed the line from then on.

When we arrived, we were driven through the gates and stopped in the coach park. Some people just sat on the grass and started eating their picnics but me and my friends wanted to explore. We found a big lake and walked around, enjoying the feeling of being away from work. I'd wanted to go into Harewood House itself, but no one else wanted to because you had to pay more, so I just had to look at it from the gardens and imagine what it must be like to live there.

We went to the bird gardens and looked round at the penguins and owls. One of my friends was paranoid about missing the coach back. She kept checking her watch and saying we'd better get back because the driver had said that he'd leave anyone who wasn't back in time. So we bought an ice cream but hung around near the coach so we wouldn't miss it.

When it was time to go, three of the younger men hadn't turned up. The driver was saying that he was going and some of the others were begging him not to. He wasn't a friendly man at all and said that if he was late back to the bus station he'd make us all pay him overtime. At the last minute, when someone was just about to go and search for them, they came running up. One of them had fallen into the boating lake and he'd had to take his clothes off. He had a towel wrapped round his waist and everybody jeered at them as they got back on. It's a shame the day ended on a bit of a sour note because it had been good, really different, but it was a very quiet coach on the way back.

In the days before coaches, charabancs transported large groups of people to places outside the city. A charabanc was an open-top bus with a roof which could be pulled over if it rained. It held around twenty people and reached speeds of thirty miles per hour. Because of its higher speed compared to a horse and carriage, it could go to further-out places like Ilkley and the Yorkshire Dales.

William Edmondson was born in Hollings Road in Manningham in 1917. He remembers charabancs coming past the top of his street and everyone cheering and waving. He was nine or ten before he rode in a car and first went to Baildon Moor in his uncle's Morris.

Lister's mill ran a trip called the 'moonlight special', which took workers to see the illuminations at Morecambe. It cost ten shillings for the return rail fare and the trains left Bradford Forster Square station at six o'clock at night. Return trains would come back as late as two o'clock in the morning.

Most trips connected to the mill were arranged on an informal basis by staff, as opposed to management. Someone would pin up a notice on the board with the price and details of the trip and, if enough people signed up, they could hire a bus and the trip would go ahead.

## Mass exodus

Bradford's geographical position also put its citizens in a fortunate position when it came to seaside holidays. Because Bradford is more or less halfway between the east and west coasts, travelling to the seaside involved a journey of just a few hours in either direction. The Edwardian period had seen the seaside holiday become popular and trains and buses left Bradford for the coast throughout the year. The mill holiday periods would be particularly frantic as hundreds of people tried to leave Bradford at once.

Jean Mortimer remembers that, on the first Saturday of Bowling tide, there was a mass exodus to the seaside. Those lucky enough to be going on holiday would find themselves waiting in a queue just to get into the railway station. She says that, although it was crowded, there was a good-natured atmosphere and the station would be full of people to catch a train as well as the people who had come to wave them off.

England's east coast offered several traditional seaside resorts, including Scarborough, Filey and Whitby, with Skegness and Cleethorpes further south, but still within easy reach. On the opposite side of the country, people would travel to Blackpool or Morecambe, and those wishing to venture further afield could access North Wales by travelling to Chester.

Diana Roberts remembers that during the summer holiday weeks there were excellent views of Bradford and beyond. This was because the mills weren't emitting their usual fumes. Diana didn't visit the seaside until she was eleven and remembers that the streets could sometimes seem quiet during the two holiday weeks in August:

I felt quite lost and lonely at times. A lot of my friends were away, some just for a couple of days; while some had a whole week at the seaside. We just couldn't afford it. I knew my mum and dad had explained this to me but at that age you just want to be like your friends. I trailed around, telling my mum I was bored and probably driving her mad. In the end, she promised that we'd go to Blackpool for the illuminations in September. I'd heard all about the illuminations from a couple of pals at school. They were on until November and it was magical to think we'd be going.

When I reminded mum about the day out a while ago, she told me she sold a bracelet to pay for that day trip, which I didn't realise. You don't as kids, do you? All I knew was that we had a brilliant time, taking the train from Bradford straight through to Blackpool and playing on the sands all day. Then, as it got dark, the illuminations lit up and it was like a fairyland. There were different themes to the lights, some were cartoon characters; others were moving people. We hadn't seen anything like it in Bradford, not even at Christmas.

It felt very late when we caught the train back; it must have been about nine o'clock at night. When we got back to Bradford, there was no taxi for us, we couldn't have afforded it, so it was quite a hike back to Listerhills.

Jean Mortimer's family went on holiday for a fortnight every year, no matter how little money they had. Sometimes they went by train from Exchange station to Cleethorpes. There would be a group of around ten or fifteen people. They spent a lot of time on the beach with their deckchairs arranged in a circle and the children playing in the centre. Jean's mother spent months beforehand making them new clothes. She remembers families would buy a pier ticket for the week, which entitled them to all the entertainment on the pier including dances.

Howard Rudge's memories of holidaying in a seaside boarding house seem a world away from the friendly guesthouses and bed-and-breakfast establishments that the Yorkshire coast is famous for today:

When you went on holiday, most people would stay in a boarding house. It wasn't like a hotel; it was literally someone's house that you were staying in. And let me tell you, you were made to feel very much an intrusion. Sometimes I can't believe we paid for the privilege of being treated how we were!

On arrival, the landlady would take you through the house rules before you were shown to your room. These ladies were quite formidable. They always wore a long flowery housecoat and never seemed to leave the boarding house. You would be allocated a cupboard and that's where you kept the food you'd brought with you. The landlady would cook it for you, but you were expected to provide the meat, vegetables, everything.

I remember my mum saying it wasn't much of a holiday by the time she'd trailed round the market getting everything for the day's meals. There were also rules about what time you had to be in at night and when you were allowed to use the bathroom. To think we thought the bathroom rules at Holden's mill were strict, they had nothing on those Scarborough land-ladies. If you missed your slot in the bathroom in the morning, the breakfast would still be put out on the table and, by the time you got down to it, would be clap-cold, they wouldn't put it in the oven or anything. But no one dared to complain, it was just how things were and those landladies were a tough breed.

For those lucky enough to be able to afford a holiday abroad, the expansion of Yeadon airport – now known as Leeds-Bradford airport – in 1953 opened up the possibility of foreign travel without having to go to Manchester or one of the London airports. Resorts in France, Spain and the rest of the Mediterranean became popular destinations.

Kim Andrews was quite shocked the first time her husband suggested a holiday to Spain, as she explained:

I'd been brought up on Butlin's holidays and had loved them from childhood. But Geoff wanted to try a holiday in Spain and he said there was no reason not to; now Yeadon airport did flights there. It was before we had the children, otherwise we just couldn't have considered the expense.

I must admit, I quite enjoyed the feeling of telling everyone at work I was going to Spain. It was 1962, I think, and going abroad was still quite a luxury thing then. I boasted that I'd send a postcard to the mill for them all to see when they got back from the fortnight off.

Anyway we had a great time apart from Geoff getting badly sunburnt. We felt like right jetsetters when we queued up to go on the plane and it was brilliant experiencing that first flight. Most people were going abroad for the first time so there was a great atmosphere on the plane and we teamed up with a couple we met who were in our resort.

When the holiday ended, I couldn't wait to get back and show everyone my tan. To my delight, my pale overalls made me look even browner and I enjoyed all the attention. But that postcard never did turn up and I got teased for weeks, with people saying I'd just put gravy browning all over myself and not gone anywhere at all.

# 11

# newcomers

Bradford's first wave of immigration occurred during the nine-teenth century and coincided with the town's rapid growth during the Industrial Revolution. While some of the newcom-ers came from abroad, the majority were from the rural parts of England and, of course, from Ireland, where the famine of the 1840s forced many to seek a new life elsewhere.

The first significant influx from outside these shores took place after the second world war. The new arrivals hailed from Eastern European countries such as Poland and Yugoslavia and, by 1955, some 8,000 Polish and Ukrainian families, together with 12,000 from Yugoslavia, had made new lives in Bradford. Many of the Eastern Europeans were 'displaced persons'. They had supported the Allies during the war, but were unable to remain in their own countries because they faced hostility from the new communist regimes. Many of them were keen to take advantage of the education and employment offered in this coun-try, while trying to preserve their native language and culture.

The next group of arrivals came from Asia during the late 1950s. The first Asian immigrants were predominantly male and they came to Bradford hoping to support family back in India and Pakistan, or to save enough money to pay for loved ones to come to Britain. By the end of our period, there were over 50,000 people of Asian origin living in the city.

Immigration can be seen both from the viewpoint of the new arrivals and from the perspective of native Bradfordians. Of course, there have been many incidences of racial tension but many of the memories related here are positive. Bradford is portrayed as a good, and welcoming, place to live. An important theme was that, after initial reservations, people were curious about workers who had arrived from overseas and wanted to make them feel welcome, as well as to learn more about their homeland. This was a unique period in Bradford's history and it helped make the city what it is today.

## Immigrants: first impressions

Living conditions were poor for workers from the Indian sub-continent. Most of them, in the 1950s at least, were single men who often had to share an overcrowded room in a boarding house. There are tales of people sharing one bed over two shifts in a twenty-four-hour period. One worker would get out of bed at six in the morning for the day shift, and a colleague who had just finished the night shift would go into the same bed.

Saiyid Abid came to the United Kingdom from Pakistan at this time. He remembers that he decided to emigrate after friends and relatives told him what a great place it was to live:

> I was working in Pakistan, quite a good job I had, but all the time when we visited friends and relations, someone would be saying that so and so had moved to England and weren't they doing well? After a while, my wife and me started to discuss it in a serious way. Yes, we realised that it would be hard to uproot the children but we saw so many newsreels about England and what a wonderful place it was. I remember seeing a film of the Queen and Prince Philip and thinking how proud English people must be to have a royal family. And it tempted me to imagine I could be part of England too.
>
> I'd been told I'd be able to find a job in a mill straight away and that although the hours would be

long, the pay would be good. Now, I didn't want to uproot my wife and two children for something that wouldn't work out, where they'd be miserable, so I decided to do what a lot of people were doing at the time, and that was to go along myself with a group of other men, find jobs, try and settle and see how it worked out. It was a long journey from Pakistan and it was unthinkable for me to put my family in a situation where they went through the trauma of all that travel for it to come to nothing.

Saima and Josinder were just three and seven years' old when I went and leaving them was a bad moment in my life. How could they understand what I was doing? And what if one of them fell ill while I was away? There were so many questions going round in my mind when I arrived in Bradford feeling quite nervous.

On the journey over, all of us immigrants discussed the fact that some of the English people wouldn't be happy to see us. You see, we weren't naïve about this at all, we didn't expect a rapturous welcome. We just wanted to do a job and send money home but we weren't stupid, we knew not everyone would welcome us. Some cousins had told me about Asians who had been spat at in the street by white people, everyone knew someone who'd experienced racism of one sort or another. It was scary to think what might happen. I tried not to think of the children because I would have just buried myself in homesickness.

You know, the first thing I remember as I stepped off that bus in Bradford was the awful smell of what I later realised was a combination of wet wool and coal. It made me wonder how I'd ever stand it. But after a few weeks I didn't even notice it anymore.

I found the first few months lonely. There were me and four other men staying in a bedsit on Girlington Road; we just went back there after work, we didn't have any interests, we didn't know what else to do. There was no established immigrant community that we knew of. I didn't know of any restaurants or cafés where I thought we'd have been welcome so we just went home at the end of the working day, it was miserable. Of course now, Bradford is full of shops and restaurants that cater for all sections of the community, but you have to know there was nothing like that it in the beginning.

I got a job easily at Illingworth's mill. I didn't even do an interview, I just met the overlooker, shook hands and I was in. Although most of the workers were fine to me, I wasn't at all fluent in English then and I didn't know a lot of the time about the films and television programmes they were talking about. But I took to the job quite easily and once I got quicker at English I could joke with the others and it was better to fit in.

I found people curious rather than hostile. They asked where I came from and why I'd come to England. I'd sit with them at mealtimes and they'd often ask me about life back home. In some ways, I just wanted to have a general chat because it hurt to think of Pakistan, but they were trying to include me and I understood that. Particularly some of the older women seemed to realize how hard it was for me leaving my family and they'd ask to see photos of the children.

After five months I decided to send for the family. It was after lots of thought and a particularly long phone conversation with my wife Shaida lasting over an hour. I stood in a phone box all the time, discussing the pros and cons with her and at last she said she'd bring the children over if I could find somewhere for us all to live. She worried endlessly about the journey, how we would meet up and how she would know when she'd got to Bradford. I wanted to go over to Pakistan and escort them over myself but I couldn't because of the job.

Knowing I was coming out of the bedsit was a big relief but I was at a loss where to find a house. I nervously approached one of the women in the canteen at work. She'd been kind before so I hoped I could ask her advice. She told me to look in newsagent's windows; they had cards about people offering places to rent. And that's how I came to get our house on Ellercroft Road, in Lidget Green.

It was about a ten-minute walk from there to the mill and there was a good parade of shops that I thought Shaida would like. Already quite a few Asian families had settled in Lidget Green and I was pleased because I wanted us to be in an area with both British and Asian families, I thought it would be good for the children.

Saiyid didn't encounter much racism himself but friends told him that people had shouted at them in the street or refused to serve them in shops. He felt proud as his children grew up with a good view of both cultures. Although they mixed with mainly Asian children, he always encouraged them to make friends with people of all races:

One thing that surprises some people is that I didn't want them to follow me into the mill. In fact, to our relatives in Pakistan, it was something they couldn't understand. Because it's usual for a son to follow his father into whatever trade. I can't complain about the textile trade, because it got me over here and into the British way of life. I could see though that it didn't have a long future.

Fahim Masood had a less rosy view of mill life, and faced antagonism as soon as he started work with a group of ten other immigrant workers in 1952:

The work was explained to us, but this phrase kept being mentioned: 'piece work'. Little did I know how much trouble it was going to cause. I just wanted to work hard to get good money but it soon caused bad feeling. There were bad looks being directed at the new workers. One of the older men came walking over and told us to slow down. 'We'll all lose our jobs the way you're going on, showing us up,' he said. None of us replied, but after that we had to be careful to keep pace with everyone else. Of course I felt angry I could have been earning more wages, but it was more important to be in a good environment.

One of the jobs often allocated to male immigrants was an unpleasant one that many British mill workers didn't want to do: washing raw wool. It involved working in very dirty and smelly conditions. Raw wool, as well as smelling terrible, could harbour diseases including the anthrax virus and pests such as mites.

A common theme is that it didn't matter what job a person had in their homeland. In England, they would have to take whatever work was on offer. Marya Krol came to Britain from Poland just after the second world war, at the age of sixteen:

I see that period as the time my childhood ended. Before the war, my father had been a doctor and he was respected in our home village. I had a privileged child-hood and never did a day's work. But one day father said we were moving to England for a better life for the whole family.

When we arrived, I didn't see how it could be a better life. We'd left behind my smart, painted village and were here in a big city that looked all grey and dull. Even the people wore clothes in boring colours and seemed to me to be so miserable. But, of course, they'd just been through a hard war like we had. I was aston-ished when my mother told me that daddy would be working driving a bus. In Poland, that wasn't a presti-gious job. How on earth were we better like this? What a selfish teenager I was!

And as for me, I had to find work and a neighbour on the narrow street who I also hated – I hated every-thing then – suggested I go to Lister's mill. It was in the Manningham district where we lived. I was shaking as mother walked me up Oak Lane towards the mill. There it was at the top of the hill, I'd never seen any-thing like it. It was storeys and storeys high; with so many windows you could never count them.

And the smoke, you can't imagine how much smoke was coming out of that chimney. You could almost see all the soot settling on the dirty buildings around. 'Tell them you had good schooling,' urged mother. Even then, I knew this wouldn't matter but I told her I would and she waited outside. I was started in weaving, just doing errands at first. You might think that your workplace would be a good place to learn your new language, but that wasn't the case at all. You see, it was so loud in those huge weaving sheds that you never heard a person talk all day.

As for me, in the end I got a job in the shipping office, arranging imports and exports. Mother was delighted; she saw it as a step up from the mill floor. But, you know, I never had as much fun up there as I did weaving. There wasn't the teamwork or togetherness.

No matter how difficult it had been to settle, many immigrants were keen to paint a good picture of their new lives to those in their native country. One good way to do this was to have a photograph taken of the whole family wearing English clothes. Bradford had many photographic studios, particularly before home cameras became common, and prints could be collected a few days after the photographs had been taken.

Nasreen Hasaan remembers her pride and excitement when her grandmother paid a visit to the family's Bradford home from India:

I was so keen to show her all the town, by then we'd been living there a year and I felt a real part of it all. We drove her down Manningham Lane one evening, to show her the Christmas lights. But she was really confused by all she saw. It must have been about 1968 and there were women walking around in mini skirts, people spilling out of pubs and nightclubs. She'd never seen anything like it. When she saw a couple holding hands in the street she was so shocked. And her reaction made me realise how much we now took for granted.

## Loneliness

Amel Khan came to Bradford from Pakistan, and remembers that she often worried about the welfare of her husband and children:

Sometimes I'd go into work unhappy as I was wondering how the children were getting on at school and whether they were being bullied. My son was being followed back home after class with people calling him names and stealing his bag and his lunch. This went on for a few weeks and I didn't tell anyone, there was no one I knew well enough. There were other Pakistani families in our area, but they had worries of their own and I didn't want to confide in them.

I didn't even tell my husband because I knew he was doing long hours setting up his own taxi business and I didn't want to put pressure on him. One day it all got too much and one of the women I worked with found me crying in the toilets. I'd just dropped Salim off at school and he'd begged me to let him come home. I told this lady what had happened and she said one of the others had a son at the same school as mine and she'd ask him to call round at the house. That evening, I really wondered if I'd done the right thing, Salim was very nervous. This boy came round that night and had a game of cricket with Salim and he'd brought some of his friends as well. When they saw how good my son was at cricket, they kept calling for him. He still got bullied at school, but having these friends outside of school helped his confidence.

Fahim Masood found that he would be most homesick at holiday and festival times:

This was particularly the case before we had good immigrant communities with things like temples and schools. For example, it would be *Eid al Adha,* the end of a period of fasting and in Pakistan we would have had celebrations with fireworks and parties lasting for days. But in Bradford, it would seem so flat. Yes, you could invite a few friends around, but we didn't make a big thing of it, maybe we were trying to fit in and not seem so different.

Now, as my grandchildren are growing up, I can see things are better in Bradford. The whole community knows about our festivals, they're taught about them at school, it's in local newspapers, I find it great. The city has made a big effort to include everyone.

## Fitting in

One of the biggest challenges facing people in a new country is trying to fit into the existing community, while retaining their identity. The desire to fit in with co-workers often led to reluctance to move further up the career ladder. This was often due to a sense of inferiority and of not wanting to leave fellow immigrants behind. Saiyid Abid remembers that he was offered promotion in a wool warehouse, but decided not to take the job. Firstly, he felt that he would be betraying the people he was working with by moving above them. He was also worried that his language skills would let him down, no matter what his bosses thought. He decided not to tell his wife that he had been offered promotion, feeling that she would be angry he had let the chance pass him by.

Polish and Ukrainian people were renowned for their enthusiasm for education. Many would encourage their sons and daughters to perform better than English children. It was a matter of family pride to tell friends and relatives that their children were doing well. Marya Krol may have had to take a mill job when she entered England, but she was still expected to attend Polish school at weekends:

I used to moan about it, but secretly I felt proud to attend. The school was in a huge house off Manningham Lane. It belonged to one of the founder members of the Polish community in Bradford. We went every Saturday, children of all ages, and I used to help teach the little ones, I liked that. We taught them about their language and customs, and we'd have dances and parties, just like in Poland. We wore our native dress, such pretty colours and ribbons in our hair. Sometimes afterwards I'd cry because it was like being back at home and I thought of my grandparents back in Poland.

Most people feel that the younger someone was, the easier it was for them to settle. Perhaps older people had more memories of their homeland, or were more aware of their culture and customs. But they were usually proud to see their children adopting Western customs, as long as they kept an awareness of their own identity.

Amel Khan remembers feeling proud as her husband's taxi business began to take off and the family were able to afford more than just the basics:

At first, it was hard. We had just enough money for food, material for clothing and our rent and bills. It was no way to live. I was constantly saying no, no, no, everytime the children asked for something. We were unusual in Lidget Green, as an Asian family having a car, but it was only because we had the taxis to use.

Gradually we got a television, central heating, telephone. These were all things that we saw on films at the Asian cinema we went to every week. And, of course, when the children saw things like that, they'd ask for them. But it was nice when they could go to school and tell their friends that they'd seen a certain programme on television the previous evening. It was all about having something to talk about and bonding, so I didn't mind it.

One thing I did worry about though was the family losing its Asian identity. As it turns out, I needn't have worried. My grown-up daughter is a stricter Muslim than me, and my son still shows his face now and again at the mosque. But, at one time, it seemed like they were so fascinated by everything the English did and wanted to copy them. 'Not everything English is great' I used to say. But they didn't listen and would carry on using English slang, wearing trainers and talking about football.

## Culture clash

Marya Krol remembers that she adopted a different personality, depending on who she was with. At school, she would try desperately to integrate with her English friends, even ignoring immigrant children from different countries. She didn't want to be associated with other strangers; she wanted to be in the gangs with English children. And then, when she went home, she would become less English, becoming absorbed again in her own culture.

She remembers feeling frustrated that her mother relied on her to go to local shops, because Marya's English was better:

I didn't want to invite any friends back to the house because I thought they wouldn't understand my mum. I'd found it quite easy to pick up English, but of course I was speaking it every day at school. And to be fair, my mum would always encourage me to show off what I'd learnt at school that day; she was genuinely interested in what I was learning.

But I felt frustrated that she didn't seem to make an effort to mix in. When I asked her what she'd been up to all day, she'd never gone out. Now I look back I can see that she must have felt quite isolated and dismayed to be alone all day in the house while her husband and children were out at work and school. But, to me, it just seemed like she was so different to my new friends' mums who went out to work in offices and shops and had lots to talk about.

Nasreen Hasaan's main frustration was her father's refusal to let her wear make-up:

All my English friends were jealous that I had my nose pierced, they thought it was so exotic and different, but I just felt angry that I couldn't wear make-up or go into town with my friends to the discos. Dad used to say that I could go to temple events and that was it. And the more I mixed in at work, the harder it got. I was upset to see that people stopped asking me along to things and I even heard some girls saying I wasn't interested in going out when I would have loved to have gone along.

The last straw was when dad told I couldn't go to the work's Christmas party. For me that was the end, I knew that no one would even bother with me after that. So I told mum that I wouldn't go out to work anymore, that's how strongly I felt and she talked dad round to the extent that I was allowed to go, but only for a few hours and only if I took one of my cousins as an escort. Not a perfect solution for me, but better than the humiliation of not going at all.

Khalida Wazir spent most of her childhood in Bradford, having been brought to the city from India in 1962. Her own family were fairly liberated and she was allowed to mix freely with friends of all nationalities. She watched many of her friends undertake arranged marriages and felt thankful when Tahir, the man she fell in love with, met with the approval of her parents. She thought she was all set for a happy life in her new house in Odsal, but little did she realise her troubles were about to begin:

I knew that Tahir's mother would be living with us. She was a widower and I didn't even question the fact that she'd be with us. Up until then we'd got on fine, but, before long, I realised she didn't like another woman in Tahir's life and she made things awkward from the start. We had our own bedroom, but that was our only privacy. And we were expected to all sit round in the front room in the evenings, not go up to our room. Tahir expected me to take his mother's advice and accept help with the cooking, cleaning and so on.

I soon discovered that my husband's mother had done everything for him and so this was a big shock because they expected me to take over where she'd left off. My own mum and dad both did things in the house, but it was different here. And I hated it that my mother-in-law kept asking when we were going to have children. We were quite happy to wait a few years and get some money behind us. But she kept saying how lonely she was when I was out at work, and how lovely it would be to have some grandchildren to care for.

I was very lucky in that Tahir's elder brother offered her a home after a year or so. She was so excited, saying how she was going to live in a big house and would be treated so well. I could have got angry but I was glad she was going. The whole experience made me more determined not to repeat this in my own family. When my two sons got girlfriends I welcomed them to the house but I kept out of arguments and made it clear I wouldn't expect to live with them, not ever.

## Bradfordians: first impressions

Just as immigrants had to adjust to a new way of life, there were also adjustments to be made by existing residents. Like any new relationship there were good and bad times but Bradford has always had a good reputation for welcoming those from overseas. This is shown by the thriving communities that exist in the city today, where ethnic shops, restaurants and places of worship are enthusiastically supported by people of all races.

On the shop floor, workers were always anxious for new-comers to settle into the job, if only for the sake of a peaceful working environment. Whenever a new wave of arrivals came on board, the mood was one of curiosity rather than hostility.

Joan Holmes had first-hand experience of welcoming a vis-itor from overseas when her family took in a lodger who had left his family behind in Poland:

I remember really clearly dad saying that he didn't want a stranger in the house but mum said we didn't have any choice as we needed the money. Not a great start, you might think, but it worked out fine. This man was really polite when he arrived and showed us pictures of his wife and two children back in Poland. I think that made my dad realise just what this chap was going through.

He stayed in our little box-room and ate all his meals with us. He always helped mum wash-up afterwards which made him popular with us kids as that had been our job up until then. I think he must have stayed a few months before he called his family to come over and they all rented a house a few streets away. I remember he brought his family around, looking so proud. His wife was really grateful to mum for taking good care of him and she used to bring round pots of soup and stew.

Kenneth Nelson's strongest memory of a group of Pakistani families moving to his area was his amazement that the children weren't allowed to play outside. He remembers seeing them looking out at the other kids playing football, but not coming out to join in. Only when they had lived in the area for a few weeks were they allowed outside as the confidence of their parents grew.

At Tyersal Combing, David Briggs was surprised the first time he saw an Asian worker carrying a rolled piece of fabric under his arm. It turned out to be a prayer mat and he remembers the Asian workers would pray in a room set aside for this purpose. Sometimes a section of the machine room would go quiet at a particular time as the Asians turned off their machines and began to pray. He also remembers Asian people bringing in delicious curries, which seemed much more appetising than the sandwiches he brought in his packed lunch. The immigrant workers were happy to share their food, and English workers enjoyed the new cuisine.

Kathleen Shuttleworth remembers that, after the second world war, many workers arrived from Eastern Europe. She says the Polish women had the reputation of being physically strong. They also stood out from the crowd as they wore ankle socks, which were never seen in the mills up until then. And she has a distinct memory of one lady who had a tattoo from a concentration camp. The Eastern European workers were also renowned for bringing plenty of alcoholic drinks to the parties.

The final word goes to Hilary Simpson, who trained many new recruits from overseas:

> Yes, we did used to get irritated with the language barrier and the fact that there were so many new people to train. But it was always just frustration with the job, never the workers who were learning it. We were all in the same boat, after all. In fact, we used to be more unkind to the women who were bussed in from Barnsley when we were short of staff.

# 12

# Decline

Bradford's output of textiles was lower in the twentieth century than it had been during the heyday of the Industrial Revolution. But this doesn't mean that the industry was in terminal decline: there were many times, particularly during the 1930s and 1940s, when the industry in the area was buoyant.

But there were also periods of great hardship. The depression of the late 1920s, linked to the Wall Street crash of 1929, brought difficult times in its wake. Many were out of work and in need of support, sometimes for the first time in their lives. The second world war brought a revival in demand for textiles and, for those working in the mills, there was the opportunity to earn good money on government contracts.

Many workers attribute the eventual decline in the textile industry to two main causes: increased competition from overseas, particularly after the war; and a change in the public's buying habits. Leisure wear became more popular and items that had been the staple product of many Bradford mills, such as raincoats and fine suits, were no longer in such demand.

## Wartime

Like every other British city, Bradford was subject to bombing during the war: it suffered five air raids, the worst in August 1940, when over one hundred people were injured. Although it was not such an obvious target as port cities like Liverpool or Swansea, Bradford could still be a dangerous place and many of the city's children were evacuated to towns like Ilkley and Hebden Bridge.

Roy Conway's father worked as a member of the Salt's mill fire brigade before he was called up for active service. When the air-raid sirens went, he had to go up onto the roof and patrol the roof gangways looking for incendiary bombs. He remembers that the rest of the family would be in the air-raid shelters while he was out. When he got back home, he would say 'the mill's still standing'.

For Joan Holmes, the clearest memory of the mill in wartime was seeing the old faces she knew so well disappear:

Before the war broke out, I'd say we were about half-and-half male and female. But soon, all the single men were called up, leaving lots of jobs free. Then some married men followed and then single girls. So the mill was made up of mainly married women.

After that, they were desperate for people to fill the jobs. We were making khaki blue, which was used for military uniforms, and with all the people being called up, it was impossible for the mill to keep up with the demand. The overlooker gathered us all round one day and said that we should think of our friends and families to see if there was anyone else who might like to come and work with us. They'd be taken on straight away. He even said the mill would be flexible about working hours. That made us laugh, they never had been before.

I'd say morale was fairly good, we felt we were helping with the war effort, even if we weren't there on the front line fighting. But, you know, there'd be really sad times too. Sometimes someone wouldn't show up at work and the next day you'd find out their husband had died or whatever. And if someone had a relative missing in action, they'd still come into work and carry on, but then break down in tears.

I think for many of us, the routine of going in and seeing our friends had a kind of normality to it. And if you were eating in the canteen, you didn't have to use your ration coupons, so you could have a really good feed and save the coupons for everyone at home.

It was a well-kept secret during the war but, as well as producing fabrics, Lister's also produced parachutes and huge pieces of camouflage cloth. These were made to look like tanks and were intended to deceive the enemy into thinking that the army had more tanks than it actually did.

Kenneth Nelson notes that, when the war was over, and people began to return to normal life, there was a marked reluctance to go back to mill work. 'It was as if the war years had changed people; inevitably I suppose,' he said. 'Most people had experienced something different during the war and didn't want to return to the bad conditions and long hours of mill work. Perhaps that was part of the reason a lot of the trade went overseas.'

## Strikes

Considering the size of Bradford's workforce, there were comparatively few strikes and industrial disputes. There were many textile unions but the industry never had a strong reputation as a union hotbed compared to, say, mining. Overlookers had to be members of the Society of Overlookers in Bradford in order to get work. Although this didn't apply to ordinary mill workers, some did join a union.

Membership of a guild or association – such as the Association of Power-loom Weavers – was another way to show professionalism and commitment to a job. Although many of them were for employers and management, there were associations that welcomed people from every part of the industry. The get-togethers hosted by these groups offered people an opportunity to network and exchange ideas about the textile trade.

In many cases, disputes were resolved without a strike. Roy Conway's mother, who was born in 1900, worked in the spinning department of Salt's known as 'the lobby'. This area was one of the biggest, and longest, workplaces in the textile world. Roy says that his mother told him that the overlooker and managers were always asking the workers to mind more machines without any extra pay. This caused resentment and, eventually, the spinners protested and switched off their spinning frames until the problem was resolved. He doesn't recall there being unions at this time but believes that solidarity paid off for the workers on this occasion.

There was a strike at Denby's mill in 1963, and a protest march from Shipley to Bradford by its workers. The strike was one of the most significant, as Denby's employed hundreds of local people. Kim Andrews remembers gossip in other Bradford mills about what was happening during the strike:

> It was over pay I think, and it went on for well over a year. We heard that the management there wouldn't back down and locked out the workers.
>
> There was an intimidating picket line by all accounts and they had to bus in people to work there. Because no local workers wanted to do it, they ended up using long-term unemployed or people just out of prison. One of my second cousins was a van driver that had to deliver there during the strike. He said the drivers were so intimidated by the picket lines that they'd refuse to drive into Denby's and so the managers at the mill would have to agree to drive out to an arranged meeting place to collect the goods the van drivers were bringing in.

## Signing-on

Despite mill work often providing a regular income, a closure or a downturn in trade had a strong effect on the neighbourhood. It was not unusual for whole families to be employed by a local mill in various capacities. Therefore, when that source of income dried up, they would be forced to search for alternative means of employment or to sign-on as unemployed.

Molly Carter faithfully paid her unemployment stamp at Northside mills, dreading the day she was ever made redundant:

> You could pay an unemployment stamp during the times you were working in case you were ever out of work and then you would get money back. If you were in a union, they gave out sick pay and they had people who went around the houses paying it out. I remember there was a man who worked in our area taking the sick money round to people on the different streets.
>
> But if someone was off sick for a long time, eventually the payments would have to stop. And that's when the family could be in real hardship. People would offer to help, taking round cakes and casseroles, whatever they could spare for families that were really struggling. None of us had much but we would think: what if it was our family, wouldn't we want people to help us?
>
> Some of the ladies would offer to mind the kids if a father was out of work so that both he and the mother could go out to look for work. If one of them found something, at least it would help out the family a little. And, of course, you can't go round looking for a job with a couple of kids in tow. Employers knew that people had children, but taking them round with you would make you look like you were going to be unreliable.

Molly remembers that when her own mill was on short time, where there wasn't enough work for everyone, some of the workers would be sent to weave at Burton's tailors in Leeds:

> We didn't mind at all, it was a bit of an adventure and a shorter day by the time we'd been taken there on the bus. It was nearly an hour's journey and Burton's always had a reputation of being a good employer. We always used to hear talk about how they provided good lighting, canteens and medical facilities long before other places did.
>
> The only thing was, some of the girls at Burton's didn't like us coming there. Maybe they thought their jobs were under threat, I don't know. Or perhaps it was just the fact of strangers coming in. They'd make fun of our broad Bradford accents that sounded different to theirs and they'd act all superior because they knew they worked for one of the big-name mills. We'd get paid in the hand for the day's work, so we enjoyed having the money up front as well. It was something different, broke up the monotony a bit and, of course, it was better than having to go and sign-on.

When mills were on short time, and there was nothing available elsewhere, workers had to go to sign-on at the labour exchange. One of Evelyn Pearson's memories is of a bus taking the whole workforce during the early 1960s:

> It was when there was a lot of work going overseas. You knew what was about to happen because you could see there wasn't the volume of orders coming through the place anymore.
>
> They took us all together to the labour exchange

and sometimes you'd meet workers from other mills. We'd all be staring at each other but not speaking. Sometimes you'd be queuing for ages in these offices. I remember one girl was pregnant and a few of us rushed ahead to get her a seat as soon as the bus stopped. We knew she'd be standing for hours otherwise.

Going to sign-on was the last resort. In the earlier part of our period, men who signed-on were given a variety of jobs to qualify for dole money. They would work a week and then have the next week off. The tasks included chopping wood and stones or working as a labourer on the roads.

When there was a heavy snowfall, such as the blizzards of 1947, there were jobs shovelling the snow. For council work like this, instead of getting paid, the money was deducted from any rates the worker owed. There was also council work available chopping wood for the town-hall fires, again on the same payment principle.

For the unemployed, there were voluntary occupation centres where people could enjoy a social life while carrying out useful occupations. There was equipment for hobbies, such as cobbling, and women took part in sewing and nursing. One such establishment was the Bradford Unemployment Advisory Centre at Manningham.

## The dreaded means test

State support was not as all-embracing as it is today. The means test was particularly dreaded. In close-knit communities, the news that someone had to go 'on the parish' (apply for help) spread quickly.

The means test involved going before a board, whose members would ask all kinds of questions about the family's circumstances and whether they had goods or furniture that could be sold. If they did, they would have to sell them and then come back when the money had run out. Diana Roberts's grandfather told her about a notorious means-test board that operated on Drummond Road:

> He told me there were five men on this panel and the person applying for the relief had to stand in front of the men, there was no chair to sit on.
>
> They'd be asked all kinds of questions about why they couldn't find work, what furniture was in the house, did they have any relatives that could help, were there any extra clothes they didn't need, that sort of thing. And only when all of those avenues had been exhausted would they make a decision. He said that in those days, which would have been in the 1920s, they had no compassion for people in need. They thought it was a form of laziness and that poor people were a drain on the city's resources.

The Manningham area benefited from soup kitchens in the 1920s. These were run from church halls or community centres and staffed by volunteers. Anyone could get a bowl of soup and whole families would visit for what, for some, was the only nourishing meal of the day. The centres would also hand out warm clothing that had been donated by members of the public.

If a family was struggling for money, their children were sometimes allowed to leave school at age thirteen to start work. All money coming in had to be strictly declared when a family was applying for relief. Joan Holmes remembers a man on her road got himself at a little job at the local newsagents, which paid a few pennies a week, cash in hand. The man ended up being jailed because he was claiming unemployment relief as well.

## Gone forever

In the early 1920s, textile production was still buoyant from the extra demand created by the Great War. The government had ordered uniforms and blankets and the resultant trade was so healthy that, in the spring of 1917, about 75 per cent of the spindles in the West Riding were used for army contracts.

The first world war disrupted exports, because Bradford wasn't able to send goods to traditional markets such as Germany and the Low Countries. Once the war was over, some of these countries either developed their own textile industries or began to trade with other countries. This was a pattern that was repeated during the second world war and was one of the main factors in the decline of the Bradford textile industry.

The euphoria created by the end of the first world war was dissipated by the economic slump of 1921, when sixty thousand people were thrown onto the dole in Bradford. The 1929 crash had an even bigger impact on with 20 per cent of people out of work in the city. During the years between 1928 and 1932, no less than 400 textile mills went out of business in Bradford.

But, for the firms that survived these difficult times, the period between the world wars was actually good for business. There was a steady demand for various textiles, as clothes became lighter and different fashions were introduced, many influenced by American films. There was also demand from stores likes Marks and Spencer. However, foreign competition was always a concern. To ease this threat, in 1931, the government imposed a 50 per cent import duty on foreign woollen and imported materials and worsted yarn. This had the effect of making home-produced textiles cheaper and caused a gap in the market that was filled by the Bradford mills.

After the second world war many textile firms struggled on but, by the 1960s, when overseas mills were equipping their factories with expensive machinery, some Bradford firms went out of business because they were either unwilling or unable to invest in the new technology.

From the mid 1950s, an influx of mainly Asian immigrants began to arrive in Bradford specifically to take jobs in the textile trade. Fewer British-born people had been prepared to return to mill jobs after the war. One of the reasons was that people became used to higher wages during the war and were not prepared to return to what they saw as poor wages and working conditions in the mill. The war had also created an expectation of a better life for those who had come through it. Parents didn't want their children subjected to the working conditions they had endured and encouraged them to pursue alternative careers.

There were new types of businesses that attracted women who would have traditionally gone into mill work. These included the Grattan mail-order company on Ingleby Road and firms that made televisions and electrical goods. These firms offered more pleasant working conditions and usually better pay.

Dee Rogers found that companies that made electrical goods were so keen to find local workers to staff their factories that they would have representatives whose job it was to hang around the mill gates when the final buzzer was due:

> For a month or so, they'd have folk waiting around the mill gates, handing out leaflets. They could never come inside the gates, but I don't think anyone could stop them approaching the workers as they left the mill.
>
> They'd have people to talk to who were already in these factories and they'd tell you what the pay was and how good the work was. Sometimes they even offered presents like a radio or record player if you joined or got a friend to join. Our bosses at the mill would try to frighten us out of going by saying that you'd get the

job and then be out of work in a month or so. 'Don't think we'll take you back', they'd threaten. But by then the writing was on the wall. Everyone knew that things were changing in Bradford and that goods like televisions, that were once luxuries, were going to be in every home.

Before a mill closed for good, there would often be large-scale redundancies in an attempt to cut costs. Elizabeth Graham said that word about redundancies always got out and everyone would be on edge:

I don't know how, but every time people were laid off, someone would find out about it before it happened and we were all on a knife edge. You'd know something was up when the machines were switched off and one or more of the directors came down. It was usually the workers who'd been there the shortest time who were made redundant first, which was only fair. But sometimes it would be the younger ones who were finished before older workers, no matter how long they'd worked there. I never understood the logic of that. Did they think it was easier for younger ones to find work, maybe? Or that younger ones might not have dependant families? I don't know, but it was hard for whoever was finished.

Usually, there'd be an announcement that there were going to be some job cuts for whatever reason. You'd never take in that part because you were just thinking 'is it going to be me?' and remembering all the bills you had to pay. And then they'd call out names to go up to the office. They'd be the people who were being laid off. They'd get a letter of recommendation, much good that would do them, and whatever pay they were due. And they had to leave the premises right away. I always found that bad, not being able to say bye or thank you.

But then sometimes a whole mill would be closing. When it happened like that, the announcement would be made, the machines shut down and everyone would just go. Some people would be kept on for a week or so, sometimes to clean up the machinery and equipment if it was going to be sold off abroad or wherever. They were grateful for any last money but it must have been a sad job after everyone had gone.

David Briggs appeared on the front page of the *Telegraph and Argus* in 2000. He was pictured under the headline, 'Just one thousand textile jobs remain', which referred to the mills left in Bradford. The article covered David's experiences of being made redundant from the textile industry on three separate occasions and finally getting so fed up that he decided to come out of textiles altogether. This was despite the fact he had greatly enjoyed the work and been part of the industry since leaving school.

The article also cited William Denby's mill, which had started on a small scale in 1820, and employed generations of Bradford people. The firm was finally going into liquidation and Briggs felt this was partly because the government had little interest in textiles. He feels the decline in the industry came about because of a lack of investment. People were training in Bradford mills and then taking their skills to other countries.

Steve Bowman, who has stayed in mill work, albeit on a freelance basis, believes that the textile industry has been squeezed because the price of clothes hasn't increased in real terms during the last twenty years. He felt that shops expected to pay the lowest price possible for the garments they were buying, even if that meant buying outside the United Kingdom.

Bowman takes the view that the way forward is for firms to cater for niche markets.

Between 1960 and 1980, over 60 per cent of the textile workforce in Bradford disappeared, with around 45,000 jobs lost. There are now estimated to be around 5,000 textile workers in Bradford compared to the 80,000 who worked in the industry in 1945.

Most former workers do not believe there was one single factor for the decline. Although the industry may have failed them in the end, for the vast majority, textiles were a major part of their lives. The mill provided them with a job; it was part of the landscape; it employed their friends and relatives. And they were proud to belong to an industry that was known worldwide for its quality fabrics.

No one can separate Bradford and textiles; the city has a proud industrial heritage and anyone who played a role is part of that history for all time. The factories may have gone but their magnificent legacy lives on. Those who played out their lives in the shadow of the mill, and who pass mill tales down through the generations, ensure that a way of life that has gone forever will never be forgotten.